Physics Research and Technology

nova
science publishers
www.novapublishers.com

Physics Research and Technology

Radiation and Matter
Marian Apostol
2022. ISBN: 979-8-88697-246-7 (Hardcover)
2022. ISBN: 979-8-88697-267-2 (eBook)

Electromagnetic Waves: Advances in Applications and Research
Manuel B. Hutchinson (Editor)
2022. ISBN: 979-8-88697-254-2 (Softcover)
2022. ISBN: 979-8-88697-265-8 (eBook)

Second Harmonic Generation: Pathways of Nonlinear Photonics
Abdel-Baset M. A. Ibrahim, PhD
Pankaj Kumar Choudhury, PhD
2022. ISBN: 978-1-68507-888-1 (Hardcover)
2022. ISBN: 979-8-88697-002-9 (eBook)

Advances in Sustainable Materials and Technology
Abhishek Kanoungo, PhD
Sandeep Singh, PhD
Er. Shristi Kanoungo, M.E
Ajay Goyal, PhD
2022. ISBN: 978-1-68507-967-3 (Hardcover)
2022. ISBN: 979-8-88697-070-8 (eBook)

Quantum Field Theory and Applications
Natale Palerma (Editor)
2022. ISBN: 978-1-68507-930-7 (Hardcover)
2022. ISBN: 978-1-68507-957-4 (eBook)

Vladimir Torchigin

The World of Circulating Light Where Ball Lightnings Live

nova science publishers

www.novapublishers.com

https://doi.org/10.52305/OMGV4259

NOTICE TO THE READER

Library of Congress Cataloging-in-Publication Data

ISBN: 979-8-88697-222-1

Published by Nova Science Publishers, Inc. † New York

Contents

Preface

The book describes a completely new world, the existence of which no one had even suspected before. Therefore, this world could not be studied either theoretically or experimentally. The study of the nature of Ball Lightning made it possible to discover this world. Pyotr Kapitsa, a Russian scientist, Nobel Prize winner, studying the phenomenon of Ball Lightning, was forced to admit at the end of his life that this plasma nut was too strong for his teeth. At the same time, he suggested that Ball Lightning is a phenomenon from another world unknown to him. Indeed, he was partly right. He was wrong that Ball Lightning was related to plasma, but he was right that Ball Lightning belongs to another world unknown to him. The book describes features of this world and the properties of its inhabitants. The most known among them are Ball Lightnings. This world we called the world of circulating light. As expected, Ball Lightnings have the neighbors, but they are also out of ordinary.

This book is not science fiction. The book is based on over eight dozen articles published since 2003 in leading international physics and optics journals. Although scientists have not been able to unravel the nature of Ball Lightning for many centuries, it turned out that the explanation is so simple that it is accessible to schoolchildren. Such an explanation can be given without a single formula, recalling only the well-known physical laws from the school curriculum.

More important is the question of whether the explanation correctly reflects the real nature of Ball Lightning? Ball lightning itself helps to answer this question with its mysterious, intriguing behavior, which is contrary to all generally accepted norm that nobody can explain.

We found an object that no one could come up with. The object is simple and unexpected, causing most scientists to stupor and instantly want to object that this cannot be for the simplest reasons. The nature of the object can be explained in a few lines. This is the usual white light from a linear lightning that has fallen into a trap that it created for himself. It rotates in this trap, gradually radiating into the surrounding space. This explains the glow of Ball Lightning.

The first simplest objection is that light propagates in a straight line, it cannot circulate. This is true if the light propagates in a homogeneous medium. In an optical fiber wound on a spool, light propagates along the optical fiber in a spiral. If the end of the spiral is connected to the beginning, then the light will circulate in the optical fiber.

While scientists shook their heads and did not believe in the existence of circulating light, we have studied the behavior of other luminous objects with anomalous properties from the same circulating world. It turned out that their anomalies can be explained on the assumption that light circulates in objects. These objects have properties that are missing from all other known objects.

Thus, explaining the nature of Ball Lightning is a long-gone stage. It turned out that Ball Lightning is one of the representatives of a wide class of objects from the world of circulating light.

No modern knowledge about quantum mechanics, the theory of relativity, nuclear physics, electrodynamics, etc. is required. Moreover, no knowledge about electricity is required either. Knowledge at the level of the 18-19th century is sufficient for understanding. Elementary knowledge of properties of light and gases is sufficient. Therefore, such an explanation could have been given at least in the 19th century.

In the Introduction, which anyone can read for free, we tried to explain what Ball Lightning is and the reasons for its abnormal behavior. We hope that after reading the Introduction, the reader will not doubt that Ball Lightning is a circulating light.

This is the only book in the world in which the correct physical nature of Ball Lightning is presented, and the physical phenomena responsible for its appearance and abnormal behavior are examined. The owners of the book will be witnesses of the fact that they lived at a time when the centuries-old mystery of Ball Lightning was solved.

The book has an appendix in which new knowledge about nature is applied to explain the Chernobyl accident. Just as the famous detective Sherlock Holmes, using his deductive method, on the basis of minor details that were left unattended by other detectives, could solve the crime, so we, on the basis of details that were not taken into account, show that the cause of the accident was the Ball Lightning that occurred in a minute before the accident.

Acknowledgments

I would like to acknowledge:

The Russian Academy of Sciences for partial financial support under project No. 0063-2019-0010 as well as the publisher, editors, and support staff at Nova Science Publishers, they were very helpful in the various stages of developing and producing the book.

Introduction

No great discovery was ever made without a bold guess.
I. Newton

Typically, the Introduction describes the state of the problem discussed in the book. The phenomenon called Ball Lightning has been known for several centuries. Currently, more than 500 different theories and hypotheses about the nature of this phenomenon are known and several thousand articles have been published. But the mysterious and intriguing behavior of Ball Lightning, which is described in many books and databases, remains unexplained. As the author of one of the scientific books about Ball Lightning said out of desperation: "All theories are united by one thing, they do not work."

The same idea can be expressed in other words. All existing theories of Ball Lightning, including the most modern ones, must be enclosed in brackets and these brackets should be multiplied by zero. Of course, this does not apply to information about numerous observations of Ball Lightning collected in books and databases. It is this information that allows us to assert that the intriguing, mysterious, paradoxical behavior of Ball Lightning in the Earth's atmosphere cannot be explained by any theory that considers some material objects.

Indeed, Ball Lightning can penetrate rooms through window panes, can find gaps and penetrate rooms through them, changing its spherical shape if necessary, can catch up with airliners and penetrate into cabins, can move against the wind, it disappears unexpectedly suddenly instantly and without a trace. Our theory explains all these mysteries and anomalies in the behavior of Ball Lightning.

Since we got rid of the need to review the existing theories about the nature of Ball Lightning, we will try to describe here what Ball Lightning is. This is not difficult to do, since the structure of Ball Lightning is very simple, and its description takes only two paragraphs.

Ball lightning is an entirely unusual physical object. Previously, no one suspected its existence. Therefore, it could not be studied either theoretically or experimentally. Of course, this object has no name.

This object is a singularity in the Earth's atmosphere that usually arises at a strike a conventional linear lightning. The strike of the conventional lightning is accompanied by the shock wave. The intense white light radiated by the edge of the shock wave occurs in a trap produced by the shock wave where hot air atoms radiate white light. In short, this object is the conventional white light circulating in a thin spherical film of conventional compressed air. This single sentence expresses the essence of Ball Lightning.

We have not met a person who would not express surprise and doubt. The light travels in a straight line. Why should it spin? If the compressed air is not confined by anything, it will expand in all directions. Why can such an object be stable? Our explanation is based on the theory of the optical model of Ball Lightning presented in eighty scientific papers in leading international scientific peer-reviewed journals in physics and optics, which are listed in the book. The articles present the reasons for the arising of stability and the conditions for its occurrence.

Any skeptic would require experimental confirmation of this model. Any model only approximately describes the real phenomenon and requires experimental confirmation. Fortunately, nature presents numerous experimental samples in the form of natural Ball Lightnings and a complete set of information about their properties in the form of numerous eyewitness accounts. From this information, it follows that the experimental sample has such a set of anomalous properties (inexplicable on the basis of modern knowledge) that are absent in numerous known models of Ball Lightning. The exception is our model, which, as shown below, has the same set of so-called anomalies. It is shown that the properties of Ball Lightning are not anomalous, but are in full accordance with the existing laws of optics and physics. Thus, circulating light is not a game of imagination, but an objective reality.

The greatest scientist Faraday after whom the unit of electrical capacity is named said: "Nothing is too wonderful to be true if it is consistent with the laws of nature." The famous American writer Mark Twain also respected reality. He believed that "fiction is obliged to stick to possibilities. Truth isn't."

The essence of our circulating light is most comfortable to explain when comparing this object with a well-known thing in the form of a soap bubble. Unlike a soap bubble, the shell of this object consists of a thin spherical layer of strongly compressed air where conventional intense white light is circulating in all possible directions. The light compresses the air to minimize the total energy of the light and compressed air. The reflective index of the compressed air is greater than that of the surrounding air. In this case, the shell is a spherical lightguide that prevents radiation of light in free space. Thus, our object is a symbiosis of two elements. These are the air and light. Compressed air helps the existence of circulating light, and circulating light ensures the existence of compressed air.

This short explanation can be clarified a bit. The air pressures inside and outside the spherical layer of compressed air are the same and equal to atmospheric pressure. Unlike a soap bubble, in which its spherical shape is provided by the excess air pressure inside the soap film, the spherical shape of our object is provided by the centrifugal force produced on the layer by the light circulating in it. The glow of Ball Lightning is explained by the fact that the circulating light is gradually scattered due to molecular light scattering. Similarly, sunlight is scattered in the Earth's atmosphere, which explains the blue color of the sky.

In these two paragraphs, we have fully described the physical nature of Ball Lightning. In the next paragraphs, we explain the reasons for its abnormal behavior. Since the time of Snell in the 18th century, it has been known that light propagating in an inhomogeneous optical medium is deflected in the direction where the reflective index increases. Thus *circulating light is moving in the direction of increasing the air reflective index.* Using more understandable physical parameters, we can say that the *circulating light moves in the direction of increasing the Earth's air density*. If we take into account that the air pressure is the same in the area where Ball Lightning is usually observed, then the lower the temperature, the higher the air density. Thus, the behavior of Ball Lightning obeys the simplest rule. *Ball Lightning tends to move in the direction of decreasing the air temperature.* However, in some cases, when moving over long distances, Ball Lightning moves in the direction of increasing air density. This situation occurs when Ball Lightning catches up with an aircraft that disturbs the uniform distribution of air density in such a way that the maximum air density is at the leading edge of the fuselage and wings of the aircraft.

Typically, the temperature distribution of the atmosphere air is influenced by various obstacles in the form of buildings, trees, etc. Besides, the circulating light itself affects the distribution of air temperature around it. When approaching such obstacles, the effect of self-action arises. Ball lightning heats obstacles with its radiation. In turn, the obstacles heat the surrounding air around them. As a result, the Ball Lightning bypasses obstacles, penetrates cracks, etc. In other words, the circulating light forces the Ball Lightning to behave entirely differently than other ordinary objects.

By analogy with the name of the soap bubble, we intended to call our object a light bubble, but confusion arose. In English, the adjective "light" has several meanings, including "light weight." Then we decided to call our object a bubble of light. However, upon further study of objects with circulating light, it turned out that they can be a ball rather than a bubble. Then we decided to simply cut off the tail in the word lightning and call our object ball light. Such a term more correctly describes the nature of these objects because there is no lightning, but there is light. Besides, the love of short words in English is well known. Thus, the term ball light or circulating light is used to designate our objects, and the term Ball Lightning is used to mean natural Ball Lightning. We will write this phenomenon in capital letters as a sign of respect for the mysterious and at the same time simple phenomenon that could not be explained for centuries

How can be proved that such an imaginary object in a form of ball light exists in nature and is Ball Lightning? The proof was carried out in the following directions.

First, knowing the physical nature of a ball light, we can show that its behavior in the Earth's atmosphere entirely coincides with the mysterious and intriguing behavior of Ball Lightning. We have deliberately chosen the names of the sections in the first chapter in such a way that by these names the reader can judge that abnormal behaviors of natural Ball Lightning coincide with the normal behaviors of ball light derived from the generally accepted laws of physics and optics.

Secondly, we have shown that Ball Lightning is not the only object in a completely new world where circulating light plays a decisive role. We present an analysis of other abnormal luminous objects from the world of circulating light and show that it is the circulating light that is responsible for their anomalies. Accounting for the presence of circulating light in these objects makes it possible to explain these anomalies.

We deliberately do not give a single formula in the book to show once again that the explanation of the anomalous properties of Ball Lightning on the basis of the well-known simple laws of optics and physics is the best evidence of the optical nature of Ball Lightning. To explain these anomalies does not require knowledge gained in the last two centuries such as quantum mechanics, nuclear physics, electrodynamics, etc. Moreover, no knowledge of electricity is required at all, since there are no electrical phenomena in Ball

Lightning. Enough knowledge of the times when Snell discovered the law of refraction of light, which is studied in elementary school.

Observational properties of Ball Lightning are given in books such authors as Singer [1], Stakhanov [2], Barry [3], Steinhoff [4], where above thousand the most revealing events are described in detail. "Ball lightning: an unsolved problem of Atmospheric Physics" - is the name of the monograph, published in 1999 [4]. We can specify this statement and say that the Ball Lightning is a shame of modern physics. Physicists cannot imagine an object obeying accepted laws of physics, the properties of which would coincide with the observed properties of Ball Lightning. Moreover, for 20 years they cannot figure out that there is a whole unique world of circulating light, in which natural Ball Lightning is only one of many other natural objects with anomalous properties.

Among these seemingly incompatible properties in one object, the following abnormal properties significantly simplify the solution of the problem.

First, we take into account the anomalous property of Ball Lightning to move in a direction that does not coincide with the direction of the wind. A separate section entitled "Flies against wind" in the Sagan book that is based on declassified Ball Lightning information of the USA Oak Ridge National laboratory [5] is devoted to a description of this property.

Second, we take into account the anomalous property of Ball Lightning to penetrate in rooms through window panes. This wonderful property is confirmed by 43 similar cases described in [6].

Third, Ball Lightning can catch up and accompany a flying airliner. An entire chapter entitled "Airplane Crashes" is devoted to a description of this phenomenon [5]. In this case, Ball Lightning is not blown away by the wind, whose speed exceeds the wind speed in the strongest hurricane.

We can conclude from these properties that Ball Lightning does not contain material objects (electrons, ions, atoms, molecules, clusters, etc.), since material things cannot penetrate through glass. Besides, since the time of Newton it has been known that the motion of any object is determined by external forces. This means that any object consisting of material particles will be immediately blown away by the wind, regardless of the specifics of internal forces and internal processes by which Ball Lightning researchers try to explain the existence of Ball Lightnings. Only light is not blown away by the wind. The light moves in accordance with its own laws.

Bearing in mind these circumstances, we can conclude that none of the known hypotheses satisfies these requirements, since it is difficult to imagine a hypothesis in which it is stated that Ball Lightning consists of nothing.

Such unusual, mysterious properties of Ball Lightning have a great advantage. They allow one to easily reject various theories and hypotheses about the nature of Ball Lightning because the objects proposed as Ball Lightning do not have the same abnormal properties that natural Ball Lightning has. On the other hand, the same set of anomalous properties allows us to assert with great confidence that an object is Ball Lightning if it fully possesses the same anomalous properties as natural Ball Lightning.

What then can an object consist of? Ball Lightning radiates light. So light is a component of the object. But how can an object radiate light if hot atoms are not included in its composition? This can only happen if the pre-produced and already existing light is somehow enclosed in the volume of Ball Lightning. In principle, this is possible if light circulates along closed paths. This is not possible in a vacuum because light travels in a straight line. In the Earth's atmosphere, which can be considered as a nonlinear optical medium, this is possible in principle.

Keeping these considerations in mind, we invented in 2003 a perfectly new object and have published the article where we have shown that our object has the same set of anomalous properties as natural Ball Lightning [7]. Properties and physical parameters of this object have been presented.

The energy density of the circulating light is greater by billion times than that of light propagating linearly. The forces arising at the interaction of extremely intense light with an inhomogeneous optical medium in the form of ordinary atmospheric air turn out to be much greater than other known forces. This fact explains all the anomalies in the behavior of Ball Lightning in the atmosphere. The behavior of well-known objects obeys the laws of mechanics. The behavior of an object consisting mainly of light obeys the same laws of mechanics. The only difference is that an additional optically induced force begins to act on objects with circulating light, which is comparable in magnitude, and sometimes exceeds other types of forces. This force also acts on ordinary objects, but its magnitude is so small that significant efforts are required to measure it, even with the current state of the art.

Since 2003 we have published about eight dozen papers in leading international journals on physics and optics where we tried to convince the scientific community that Ball Lightning is circulating light and to show that all anomalous properties of natural Ball Lightnings are identical to that of our ball light [7-88]. The present book is based on these publications. Note that the number of publications on our theory is an order of magnitude larger than on any other known theory or hypothesis. Usually, such theories are limited only to the suggestion of some luminous object, which may have one or several properties of natural Ball Lightning. Further studies of this object show that the properties of this object are far from the anomalous properties of natural Ball Lightning. We hope that the properties of circulating light presented in this book so convincingly that the reader will not doubt that circulating light is an objective reality.

The correctness of the optical nature of Ball Lightning does not follow from the theoretical reasoning but is based on the striking similarity of the behavior of the ball light and Ball Lightning in all situations in which the inexplicable, mysterious, and intriguing behavior of Ball Lightning is noted. A similarity of behaviors of ball lights and Ball Lightning is shown for two dozen of the most mysterious situations presented in the first chapter.

There is one difficulty in writing this chapter. For those readers who are familiar with the laws of optics and they have the idea of such concepts as the reflective index, density, speed, and acceleration, a consideration of the behavior of a ball light is obvious without additional explanations and the chapter can be decreased to several pages as it was done in [42].

Since our goal is to convince the largest readership of the existence of ball lights, an elementary explanation is given without any mathematics. As Newton taught, in the study of sciences, examples are more important than formulas. We hope that only this single chapter is sufficient to persuade the reader about the optical nature of the natural Ball Lightning.

The chapter could be considered as a separate book, but the circulating light manifests itself not only in natural Ball Lightning but also in other objects that have been studied for two centuries. Therefore, natural Ball Lightning is considered only as a messenger from the wider world of circulating light. Explaining the nature of Ball Lightning is only the beginning. Ball lightning and circulating light for free demonstrate a lot of new information, which is presented in the second chapter. We believe that this chapters may be additional arguments that Ball Lightning is a circulating light.

For the past two centuries, researchers have been working on producing Ball Lightning in a laboratory. They obtained luminous objects with anomalous properties. However, neither in size nor in life time, they corresponded to a natural Ball Lightning. Analyzing the anomalies of these objects, we came to the conclusion that the anomalies can be explained by assuming that the objects contain circulating light. Then one can consider a natural Ball Lightning not as a unique phenomenon, but as one of the representatives of a wider class of phenomena in which circulating light participates.

Unlike a natural Ball Lightning, which is a rare natural phenomenon and cannot be systematically investigated, the physical properties of anomalous objects can be analyzed using existing test devices. Over the past two centuries, a great deal of experimental data has been accumulated on the parameters of anomalous objects. Just by studying these data and bearing in mind the participation of circulating light in an appearance of anomalies, we can explain the nature of these anomalies and obtain a lot of valuable information from the world of circulating light at the time when sceptics shake their heads and express doubts about the optical nature of Ball Lightning.

For example, an analysis of the emission spectra of abnormal phenomena when a liquid nitrogen vapor is illuminated by a powerful flash lamp allowed us to discover a phenomenon of self-organization of intense light in gasses. Also, we discovered a new mechanism for increasing the reflective index in a gas mixture by studying the published spectrum of luminous anomalous objects. Comparing the values of the reflective index of the elements represented in the spectrum of anomalous objects, we found that there are elements, the reflective index of which is greater than that of the air. Thus, nature suggested another way to increase the reflective index in the shell of ball light. The reflective index may increase not only due to an increase in the air density because of an increase in the air pressure but also due to the drawing into the shell of the components of the gas mixture with the maximal reflective indices.

Another example. It was known from experiments that an erosive gas discharge has a beneficial effect on the production of luminous objects with anomalous properties. Taking into account the mentioned above mechanism, we enabled to explain this beneficial effect.

One more example. More recently, a video appeared on YouTube in which a luminous ball jumps on a table for 8 seconds, making over a dozen bounces. Only taking into account the circulating light made it possible to explain this phenomenon and show that circulating light can exist not only in compressed air but also in a liquid. Besides, the circulating light can arise and exist due to exothermic chemical reactions.

For almost 100 years, a phenomenon has been known when, under the action of ultrasound, a bubble begins to glow in the water. Moreover, its spectrum corresponds to a temperature of a hundred of thousand degrees. This phenomenon has received its own name sonoluminescence. Attempts are known to use this phenomenon to carry out thermonuclear reactions in a test tube with water. We have shown that the glow of the bubble is associated with circulating light, the spectrum of which shifts to the short-wavelength region when the light is compressed with decreasing the bubble size. Therefore, there is no temperature of hundred thousand degrees in the bubble, and attempts to obtain a thermonuclear reaction are in vain.

Two decades have passed since convincing evidence was presented that the behavior of the circulating light and natural Ball Lightning is the same, and therefore the circulating light exists in nature. It would seem that the discovery of a new natural phenomenon should have aroused an interest in it and in the study of the presence of the circulating light in other natural phenomena. However, this did not happen.

Bearing in mind the extremely high air compression and the extremely high energy density in the shell of the circulating light, we could not disregard the consequences of this circumstance. Many descriptions of luminous objects with anomalous properties are accompanied by a mention of the occurrence of the excess energy. Such excess energy can only be obtained from the energy of the nuclei. The third chapter presents an analysis of how unique parameters in the shell of the circulating light can affect an extraction of the energy from nuclei. Keeping in mind that the particles of the micro-world have not only corpuscular, but also wave properties, the conditions under which wave properties can be used to extract the energy from nuclei are considered. It is shown that this is possible with a long-term interaction of two identical nuclei located in closely spaced atoms. In this case, the energy from one nucleus can pass into another, which leads to a violation of stability, as a result of which a nuclear reaction occurs. A table of possible similar nuclear reactions is given, indicating the excess energy in such a reaction in the range from tens to hundreds of keV per two identical nuclei.

Having new knowledge about the properties and physical parameters of the circulating light, it is reasonable to analyze whether this knowledge can be used to explain previously occurring unusual phenomena. Currently, many countries are demonstrating operating installations that produce electricity without the use of traditional fuels and traditional energy sources. These installations extract the energy either from the air or from the water that drives these installations. Assuming that the energy required to operate such installations is extracted from the energy of the nuclei, we came to the conclusion that the energy is extracted using the circulating light that occurs in the air bubbles used in the operation of each of them.

In Chapter 4, we consider the principles of operation of such installations, give their shortcomings and ways to eliminate them. At the same time, we give a description of the design and principles of operation of alternative installations that use the same physical phenomena to extract the energy from nuclei. These installations can be used to generate electricity, as well as to propel ships.

Analyzing the new knowledge gained, we came to the conclusion that this knowledge can be used for an alternative explanation of past anomalous phenomena, in particular, for explaining the causes of the Chernobyl accident. It is shown that the real cause of the accident could be the Ball Lightning that occurred a minute before the explosion at the beginning of the scheduled harmless electrical tests. Our arguments are presented in Appendix.

It can be concluded that the book not only explains the physical phenomena responsible for the existence and anomalous behavior of a natural Ball Lightning, but also expands our knowledge of nature. It is shown that there is a hitherto unknown world of the circulating light with its own characteristics, inhabited by objects from the ordinary world, in which the properties of these objects are considered as abnormal. This book is the first book that mentions anexistence of a world of the circulating light, describes and explains the properties of various objects from this world.

Chapter 1

Evidence of Identical Behavior of Ball Lightning and Circulating Light

Truth is ever to be found in simplicity, and not in the multiplicity and confusion of things.
I. Newton

Following our ideas about Ball Lightning as the most famous representative from the world of circulating light, our first task is to convince the reader that this is indeed so. In this chapter, we use the most convincing way of persuasion, in our opinion. We show that the anomalous behavior of Ball Lightning, known from numerous eyewitness accounts, coincides with the behavior of the circulating light, which can be obtained theoretically. Indeed, since the nature of Ball Lightning in the form of a ball light is known, the laws that are responsible for its behavior are also known. Fortunately, the behavior of Ball Lightning is so exotic that not one of the other known hypotheses even tries to explain this behavior. It is easy to understand that the explanation of exoticism is more credible than the explanation of ordinary phenomena.

Based on a synthesis of numerous eyewitness accounts, English Wikipedia lists the following anomalous, mysterious, and intriguing properties of Ball Lightning:

- Ball lightnings frequently appear almost simultaneously with cloud-to-ground lightning discharge.
- Ball lightnings are generally spherical or pear-shaped with fuzzy edges
- Ball lightning diameters range from 1–100 cm, most commonly 10–20 cm
- Ball lightning brightness corresponds to roughly that of a domestic lamp so that it can be seen clearly in daylight.
- A wide range of colors has been observed, red, orange, and yellow being the most common.

- The lifetime of each event is from 1 second to over a minute with the brightness remaining fairly constant during that time.
- Ball lightnings tend to move, most often in a horizontal direction at a few meters per second, but may also move vertically, remain stationary or wander erratically.
- Many Ball Lightnings are described as having rotational motion.
- Observers rarely report the sensation of heat, although in some cases the disappearance of the ball is accompanied by the liberation of heat.
- Some Ball Lightnings display an affinity for metal objects and may move along conductors such as wires or metal fences.
- Some Ball Lightnings appear within buildings passing through closed doors and windows.
- Some Ball Lightnings have appeared within metal aircraft and have entered and left without causing damage.
- The disappearance of Ball Lightning is generally rapid and may be either silent or explosive.
- Odors resembling ozone, burning sulfur, or nitrogen oxides are often reported.

We will not comment on these properties. The reader is referred to the Sagan book [5] where systematization of the anomalous properties of Ball Lightning is given. The same book contains a list of the physical laws that are violated by Ball Lightning in Sagan's opinion. In our opinion, no laws are violated. Because of this to list these alleged violations is senseless.

There is a simple and reliable way to make sure that the optical model correctly describes the processes in Ball Lightning. It is enough to compare the behavior of Ball Lightning based on numerous eyewitness accounts presented in books on Ball Lightning and the behavior of the ball light, which can be obtained based on the generally accepted laws of physics and optics. If these behaviors are identical, then it is safe to say that the optical model of Ball Lightning is correct. None of the 500 current theories of Ball Lightning has even tried to explain the intriguing, paradoxical, and abnormal behavior of Ball Lightning.

We believe it is worthwhile first to demonstrate the similarity of the behavior of the ball light and Ball Lightning in situations in which the Ball Lightning is most often observed. This should convince a reader that the circulating light and the ball light are other names for a Ball Lightning. Two preliminary short section are necessary for this purpose. The section gives a preliminary explanation of the laws governing the movement of the circulating light in an inhomogeneous atmosphere of the Earth and presents the typical phenomena responsible for the inhomogeneity of the Earth's atmosphere. This is sufficient to explain a behavior of the circulating light in an inhomogeneous atmosphere and to compare this behavior with that of Ball Lightning.

It would be right to start the book with a statement of those laws and physical phenomena that are responsible for the existence and abnormal behavior of the circulating light. However, the specificity of this book is that it provides the information that most readers can attribute to science fiction and do not want to delve into the formulas. Therefore, we decided to start the presentation with our main argument that circulating light is not a game of imagination, but an objective reality that behaves in the Earth's atmosphere in the same way as Ball Lightning. In this chapter, you will not find a single formula.

The Motion of a Ball Light under Laws of Geometric Optics

In an inhomogeneous optical medium, a beam of light is curved in the direction where the reflective index of the medium is increased. The most famous example of curvilinear propagation of light is mirages in deserts when an observer can see objects beyond the horizon. This can take place only if the light propagates along a curve trajectory from the object to the observer's eyes, as shown in Figure 1.

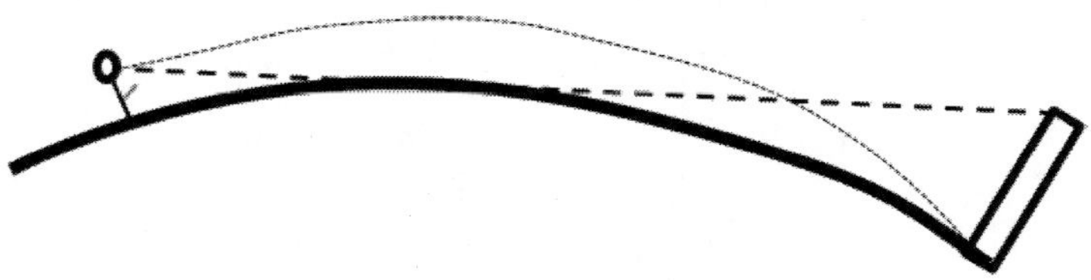

Figure 1. The top of the building is seen when a light beam propagates in the homogeneous air. The whole building is visible when the beam propagates in the air of inhomogeneous density. In this case, the light beam deviates from a straight line at a distance equal to the height of the building.

We need to explain in a popular way what are inhomogeneous optical media and reflective index. In this book, we consider the reflective index as the ratio of the speed of light in a vacuum to the speed of light in an optical medium. The greater the air density the greater the reflective index. An inhomogeneous optical medium is a transparent optical medium where the reflective index is not constant and depends on spatial coordinates. Under the basic law of geometric optics, a light beam bends in the direction of increasing the reflective index. Thus, from Figure 1, we can conclude that the reflective index increases towards the center of the Earth.

Let us estimate the beam shift in the transverse direction under the assumption that an observer sees the object at a distance of 30 km. As you can see from Wikipedia, the horizon for the observer is 4.7 km. At a distance of 30 km, the observer with a straight-line propagation of light sees the top of an object with a height of 50 m. If the observer sees the whole object in the desert, then the beam at a distance of 30 km is deflected by 50 m. The light travels a distance of 30 km behind 10^{-4} s at the speed of light of 3 10^{8} m/s. Then the beam velocity in the transverse direction is $50/10^{-4}$ = 500 km/s.

To make it clearer, we divide the beam path of 30 km in Figure 1 into segments 30 cm long. There are 10^{5} such segments. The beam shifted downward at a distance equal to $50/10^{5}$ = 0.5 mm when passing through each such segment.

The same deflection we will have if the light beam reflects periodically in the same inhomogeneous air from two ideal reflectors in the form of two parallel vertical mirrors separated at a distance of 30 cm as is shown in Figure 2a. If the optical medium between the planes is identical to that in the desert, the light beam will need to move downward at the same speed of 500 km/s, and each segment of 30 см long is shifted by 0.5 mm.

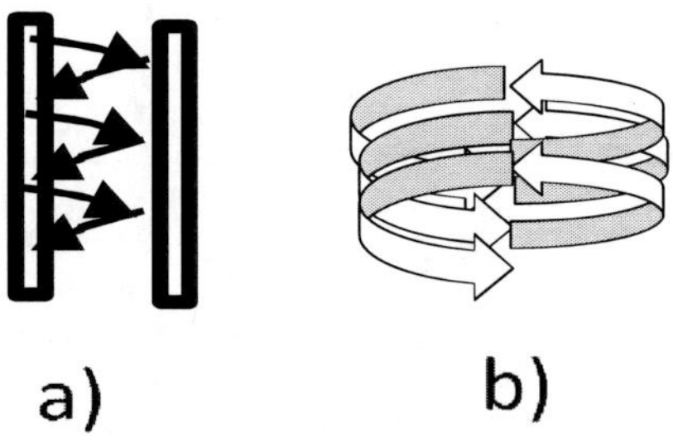

Figure 2. Comparison of shifts downward of the light beam propagating in an inhomogeneous air
a) at the reflection from two vertical parallel reflectors
b) circulating light without reflectors.

Consider now the circulating light shown in Figure 2b instead of the light between the reflectors. In this case, reflectors are not required, since, during rotation, the light itself periodically changes its direction of propagation. In this case, the light will also shift downward. We can say that the circulating light is screwed into the optical medium in the direction of increasing the reflective index. For example, if the diameter of the circle around which the light rotates is 10 cm, then the light makes about one billion revolutions per second and shifts by 500 km. Then it shifts by 0.5 mm per one revolution.

We can say that the trajectory of light deviates from a straight line under the influence of optically induced forces (OIF) (an approach based on the laws of mechanics is used instead of the approach considered here based on the laws of geometric optics). Indeed, such forces exist and provide the movement of circulating light in the direction of increasing the air density. Instead of mentioning each time that the circulating light moves in the direction in which the air density increases to the maximum extent, we will simply say that the circulating light moves in the direction of optically induced forces. In this case, we mean that the optically induced force acts in the direction in which the air density increases to the maximum extent and a magnitude of this force is proportional to the air inhomogeneity in space.

In particular, if the circulating light is located in the place of space where the air density is maximum, then any deviations from this position are accompanied by arising OIF directed towards increasing the air density, that is, towards the initial position.

Of course, the inhomogeneity of the optical medium in the form of air under ordinary conditions when Ball Lightning is observed is significantly smaller than the considered inhomogeneity of air in the desert. For example, if the circulating light moves at a speed of 5 m/s, the inhomogeneity of the medium will be 100,000 times smaller. We can say that circulating light is an extremely sensitive device that responds to the slightest inhomogeneity in the air where it is located.

First, we must convince readers that the behaviors of the circulating light and Ball Lightning in the Earth's atmosphere are the same. Since the behavior of the circulating light depends on the inhomogeneity of the air in which it is located, we first consider the causes leading to an appearance of such inhomogeneity.

However, first, it is necessary to determine an indicator characterizing the degree of air inhomogeneity. In optics, this indicator is a change in the reflective index in space. However, not everyone can imagine the change in the reflective index in the air atmosphere. The reflective index of air increases with increasing the air density. This indicator is more understandable, but it is not so easy to imagine how the air density in the air atmosphere changes. It is known that the air density is proportional to the air pressure and inversely proportional to the air temperature, expressed in degrees Kelvin. As for the air pressure that it decreases with the altitude. At the same time, the air pressure in a limited region where Ball Lightning is observed is constant.

However, the temperature in the region where a behavior of Ball Lightning is observed is not constant. It is a change in temperature that is responsible for the movement of the circulating light. Where the temperature is lower, the density of the air is greater, and the circulating light moves to this area. The fact that the density of air decreases with increasing temperature has numerous confirmations. For example, hot air from a fire rises upwards. To reduce the air density of the air in a balloon, balloonists use a burner that heats the air inside the balloon.

Thus, the behavior of the circulating light near the surface of the Earth obeys a very simple rule. *The circulating light moves in the direction of decreasing the external air temperature.* The air temperature is an understandable parameter that does not require an additional explanation. *This simplest primitive rule explains all puzzles, mysterious and intriguing behavior of ball lightnings in the air atmosphere that could not be disclosed in many centuries.*

Now we need to analyze what can affect a change in the air temperature. First of all, these are buildings. People try to live at the temperature that is comfortable for them, and not at the temperature that nature provides them. This explains the widespread use of various heaters and air conditioners. As a result, the temperature of the walls of buildings differs from the ambient temperature. This is especially noticeable near various inhomogeneities of the walls such as holes, windows, open windows.

Besides, it must be borne in mind that the heat capacity and the thermal conductivity of the surface of the Earth, air, and buildings are different. As a result, they react differently to the 24-hour cycle associated with the rotation of the Earth around its axis.

For example, buildings have a foundation buried several meters in the ground. The temperature at this depth is close to the average annual temperature and varies slightly during the day compared to changes in the temperature of the roof and walls. At the same time, due to the phenomenon of thermal conductivity, all parts of the building are interconnected.

Besides, the circulating light itself affects the temperature of the air. The air is transparent and cannot be heated by the light that radiates the circulating light due to the phenomenon of molecular light scattering. The radiated light is visible to the observer of a natural Ball Lightning. However, the situation changes if the circulating light is near some obstacle. The obstacle absorbs the radiated light, and the surface of the obstacle heats up. The heated surface heats nearby air layers, and the air density in these layers decreases. The circulating light must respond to this circumstance.

This simple and indisputable information is enough to explain all the anomalies, secrets, and intriguing behavior of natural Ball Lightning.

Comparison of Behaviors of a Ball Light and Ball Lightning in Various Situations

Why Ball Lightning Does Not Hit the Ground at Falling from the Sky

Many scientists are perplexed by the following fact. All Ball Lightnings that fall to the ground from heaven do not hit the ground, but slow down and stop at a certain distance from the surface of the Earth as if they would be afraid of breaking on the ground. This property of Ball Lightning distinguishes them from all other known objects. Whatever internal processes occur inside Ball Lightning, its behavior is determined by external forces. These forces should somehow distinguish Ball Lightning from all other known objects that, when falling from the sky, hit the ground. These forces should begin to slow down Ball Lightning so that its speed decreases to zero at a distance of several meters from the ground. Several questions immediately arise. What is the nature of these forces? Why do they begin to act only directly near the surface of the Earth? What is the difference between this distance and other distances?

It can be assumed that Ball Lightning has a jet engine, which is turned on to brake directly around the Earth. In addition to the question of how such an engine is designed, other questions arise. How does this engine work? What instrument, like an altimeter, do Ball Lightnings use to determine a distance to the Earth? What tools does the Ball Lightning include to slow down to zero and not hit the ground? It would seem impossible to find reasonable answers to these questions; however, all questions receive their answers if we admit that Ball Lightning is the circulating light.

An answer is much shorter than the questions [32]. The circulating light moves in the direction of increasing air density and stops at the height where the air density is maximum. It cannot move either up or down, as the air density decreases in these directions. As is shown in Figure 3c, the maximum air density is located at a distance of several meters from the Earth's surface.

Indeed, as the air pressure decreases with increasing the altitude as is shown in Figure 3a, the air density also decreases, and hence the gradient of the air density is directed downwards. Since a ball light moves along the gradient of the air density, the ball light falls until the gradient of the air density is directed downwards. But the ball light does not reach the Earth's surface. The reason is that immediately at the Earth's surface the temperature is greater than that at the height of several meters as is shown in Figure 3b.

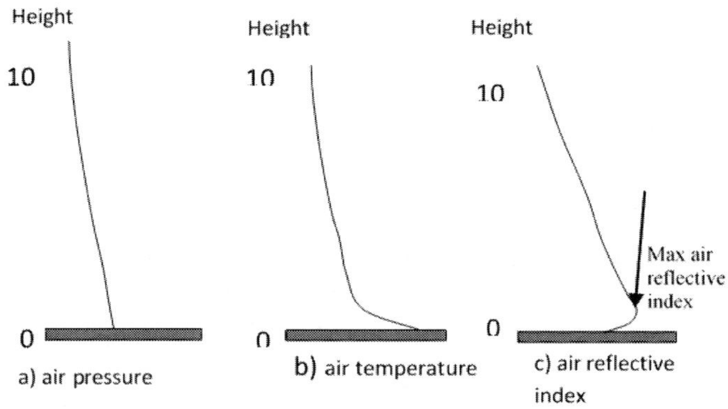

Figure 3. Dependence of the air parameters on height taking into account solar radiation.

This is due to heating the surface of the Earth by solar radiation. Immediately near the surface of the Earth, the ball light should move in a direction of decreasing temperature, that is, upward until the density of the air increases. At the same time, it reaches a height where the air density is maximum as is shown in Figure 3c. It cannot rise higher, since the air density will decrease. At this altitude, the ball light stops its movement in the vertical direction. The force of gravity acting on the ball light is not essential since the optically induced forces (OIF) exceed the force of gravity.

Stopping the vertical movement, the ball light continues to move in the horizontal direction opposite the horizontal component of the gradient of the air temperature at the height where the vertical component of the gradient of the air density is equal to zero. That is why the observed Ball Lightnings are mainly moving at low height in the horizontal direction.

We can remind the reader that the fog lamps on vehicles are located as close as possible to the surface of the Earth because the air near the surface has the maximal temperature, and the mist at the temperature is not formed. Therefore, the fog lamp shines the road to a greater distance. The air density decreases with increasing temperature and, therefore, there is a maximum of the air density at a certain height from the Earth's surface where Ball Lightnings are moving. Thus, the intriguing behavior of Ball Lightnings, giving an appearance that it does not obey the laws of gravity, is explained naturally if the optically induced forces are taken into consideration.

Why Ball Light Moves Horizontally

As you know, an ordinary children's balloon filled with helium flies up if the child releases the thread that prevents it from flying. Another ball can lie on the ground if it is filled with ordinary air. But there are no balls that reside at a certain distance from the ground and do not rise or fall. Knowledgeable people will say that two forces act on the ball; one of them is the force of gravity, while the other is the buoyant force or the force of Archimedes. If gravity is greater than the force of Archimedes, the ball will rest on the ground. If the gravity force is smaller than the Archimedes force, the ball flies up. A situation in which gravity is equal to the force of Archimedes is very rare since there is a high probability that one force prevails over the other.

Unlike a children's ball, Ball Lightning does not fly up, nor falls to the ground, but moves near the surface of the Earth.

An explanation is simple. The maximum air density (and therefore maximum air reflective index) in the vertical direction is at some distance from the Earth's surface as is shown in Figure 3c. The circulating light moves at this height.

Note that the usual children's balloon cannot move by inertia in the horizontal direction for a long time, since the initial horizontal component of velocity decreases rapidly to zero due to the Stokes force. Stokes force is responsible for the resistance experienced from the air by any moving body. As a result, the pressure on the front hemisphere of the ball light is greater than that on the back hemisphere. Then the reflective index near the front hemisphere is greater than that near the back hemisphere. Since ball light tends to move in the direction where the reflective index increases, there is an optically induced force directed opposite to the Stokes force. This force can compensate for the Stokes force, and the ball light can move horizontally as if the Stokes force is absent [34].

How Ball Light Feels Obstacles

How does Ball Lightning feel that an obstacle has appeared in its path and what means does it use so that it does not run into the obstacle, but bypasses it?

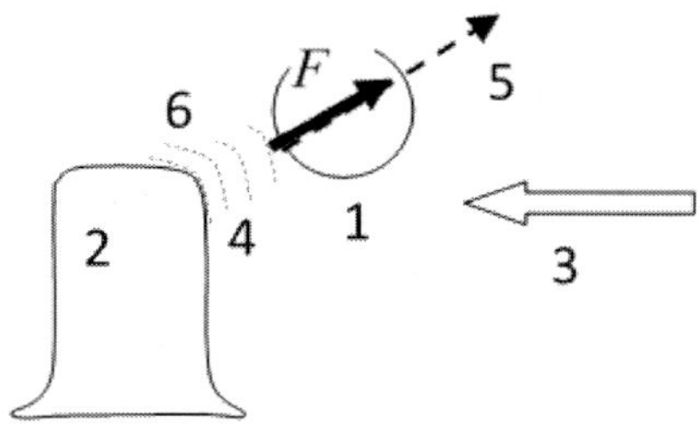

Figure 4. How ball light 1 bypasses obstacle 2 where 3 is the direction of movement, 4 is isotherms of air, 5 is the direction of maximal decrease of the air temperature, 6 is the component *F* of the force arising due to inhomogeneity of the air temperature.

As is shown in Figure 4, ball light 1, moving in the direction shown by hollow arrow 3 and approaching obstacle 2, heats the obstacle due to radiation emitted by the circulating light. The obstacle warms up nearby air layers 4. The closer to the obstacle, the more it heats the air. As a result, an inhomogeneous optical medium arises near the obstacle. The air temperature decreases along arrow 5 indicated by the dashed line in Figure 4. In this medium, OIF directed along this arrow acts on the circulating light. The smaller the distance between the ball light and the obstacle, the greater OIF. Once in this zone, the circulating light deviates in the direction of OIF, that is, from the obstacle. Outwardly, it looks as if the ball light "senses" an obstacle and avoids it.

How Ball Light Detects Holes to Penetrate through Them

It is known that Ball Lightnings penetrate various openings and crevices into the rooms. Several questions immediately arise. How does Ball Lightning find these openings? What sensory organs are used? Further, what means does Ball Lightning use to organize its movement towards the opening? It is clear that the Ball Lightning has no organs of sight, smell, touch, that have living beings.

If Ball Lightning is the circulating light, the answer to this question is simple [38].

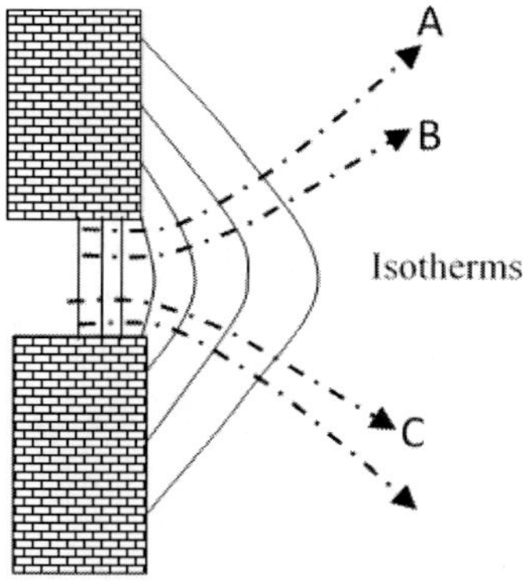

Figure 5. Surfaces of equal temperatures near the split in a wall are shown by solids. Tangents to the surface (dotted lines) show directions of the gradient of the temperature directed from the hole.

Consider a motion of the ball light near a tunnel between a room and outside space. Suppose that the room temperature is lower than outside, and the air pressures inside and outside are the same. In this case, the air temperature on the axis of the tunnel increases with distance from the room. Since the air temperature varies gradually and continuously when moving along the tunnel, the surfaces of equal temperature have the form shown in Figure 5 solid lines. Dashed lines in Figure 5 are perpendicular to these surfaces. The direction of the tangent to any such line at any point coincides with the direction of the increase of the air temperature at that point. It is easy to verify that the ball light, moving along any of the dashed lines from various points A, B, C, moves toward the tunnel. A kind of a funnel is formed near the tunnel. Once in this funnel, the ball light moves towards the tunnel.

Likewise, ball light detects rooftop chimneys to sneak through them into the room. There is a funnel around each chimney. Falling from a height, the ball light falls into one of these funnels and begins its movement into the room. If necessary, the ball light can change its shape, as is discussed below.

How Ball Light Changes Its Spherical Shape

Ball lightning can penetrate the room through the door cracks, windows, chimneys, may leave the rooms. There are numerous reports that Ball Lightning moved along the walls of the room at a certain height, avoiding the emerging obstacles in its way, comes to slits, and penetrates through them into the room. The dimensions of the slots or openings can be considerably smaller than the diameter of the Ball Lightning.

There is a huge amount of evidence that Ball Lightning can penetrate through small slits. Let us present a few of them, taken from the book of Stakhanov [2].

During a severe thunderstorm, Ball Lightning of 20-30 cm in diameter entered through a hole in the wall for grounding.

During thunderstorms, Ball Lightning 10 cm in diameter penetrated in the hole 2 cm wide. Ball lightning deformed in "stretched sausage." while penetration.

Ball lightning 10-20 cm in diameter went into the crack around the window glass.

Ball lightning diameter of 30-50 cm entered through a small hole in the window (glass chipped corner) 1-1.5 cm in width as the "yellow thread." Ball lightning exploded after 20-30 seconds.

Ball lightning 5-10 cm in diameter entered as a "snake" during a thunderstorm through the open window and then forming a bead. After going around the room at a distance of 5-10 m, Ball Lightning disappeared without an explosion near the switch.

Ball lightning 10-20 cm in diameter passed through the hole diameter of 8 cm.

Ball lightning "the size of a tennis ball" has gone through a closed window, in which the glass had a crack.

Ball lightning "flowed" in the hole between the logs in the forge room. The slit width was much smaller than the Ball Lightning diameter. Ball lightning was a ball of 12-13 cm diameter, orange, the brightness of the lamp 50-100 watts.

A yellow ball of a large orange size was creeping through the crack in the wall. Rather, it was not creeping but was poured from one-half to the other.

Ball lightning walked into the room through a hole in the glass, was flattened, as its size was larger than the hole. Eyewitnesses clarify: "the ball was in the 10-15 cm from our faces, and we have seen well as it began to pass through the hole, taking the form of a melon. It stretched out, was less in diameter, and passed through the hole. When the ball passed through the hole and decreased in size, it was shaking all the time, and it seems that it consists of jelly."

How does it recognize the shape of the hole and change its shape to match the shape of the hole?

Consider the reasons for which ball light can change its shape to penetrate the room through a small slit or hole [28]. First, we must explain the reasons for its spherical shape in a homogeneous air and its deformation in inhomogeneous air. As already mentioned, there is the sufficiently great air pressure in the thickness of the ball light shell. The spherical shape provides a minimum of compressed air energy amongst other possible shapes. However, this does not mean that Ball Lightning is like toy balloons, which is difficult to deform significantly.

The air pressure exists only in a thin ball light shell of several micrometers thickness. This pressure provides the spherical ball light shape but to a much lesser extent. The small external forces can significantly change the ball light shape. It reminds a shell of a baby balloon from which the air is removed, and the balloon is placed in a vacuum and weightlessness. The shell takes the form of a sphere under the action of its own weak elastic forces tending to straighten it and make it flat. Since the topology of the shell is such that it cannot be flat, elastic forces in the absence of other forces cause the shell to take the shape of a sphere. This shell can be easily stretched through a wedding ring.

The shell of circulating light is permeable to air. The air pressures inside and outside the shell are the same. The spherical shape of the circulating light is provided by the pressure within the thickness of the shell itself like the shell of the toy balloon, from which the air is removed, and the balloon is placed in a vacuum and weightlessness.

As is seen in Figure 6, it is obviously that the thinner the shell, the higher its deformability.

Figure 6. Comparison of deformability of two cameras of different thicknesses. The right camera is thinner than the left. There is no doubt that the thinner the layer of compressed air, the higher the deformability of the chamber.

A similar situation holds for the ball light shell, which adopts a spherical shape in a homogeneous optical medium due to the elastic properties of the layer of compressed air in the shell. The ball light form can be easily changed in the same way as the spherical shape of the elastic child balloon in a vacuum, and the weightlessness is changed.

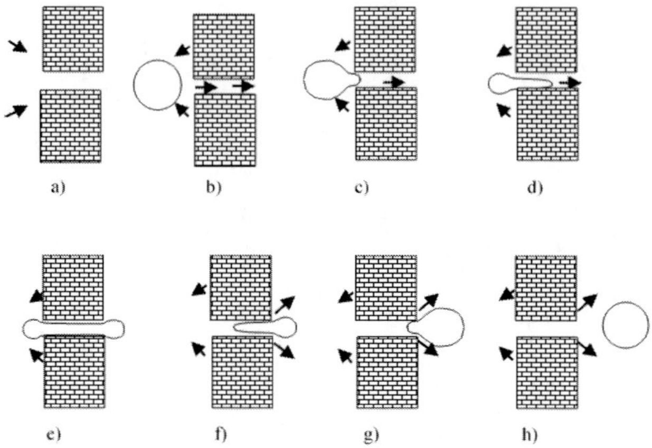

Figure 7. Sequential steps of penetration of the ball light through the split in a wall.

When the ball light is approaching the opening in the wall, wherein the gradient of the air temperature is directed from the wall, a distribution of the gradient of the air temperature shown in Figure 7 is changed due to the appearance of the following effect of self-action. When approaching the ball light to the hole, an inhomogeneity of air temperature increases because ball light heats the wall, which in turn heats the layer of the air between the ball light and the hole. Thus, ball light indirectly heats the air. Solid objects that absorb radiation radiated by ball light must first be heated. The regions surrounding the hole are used as these objects. These regions heat the adjacent air layers due to the phenomenon of heat conduction.

The closer the air layer to the opaque surface of the wall is located, the higher its temperature. OIF arises that repels the ball light regions, which are closer to the wall, from the wall surface. At the same time, according to our initial assumption, the air temperature gradient directed from the room continues to exist. As a result, optically induced forces of different directions appear, applied to different parts of the shell of the light ball. The action of these forces leads to the deformation of the light ball since the restoring properties of the light shell of the ball are extremely weak. The ball light takes the shape shown in Figure 7c.

As the ball light penetrates the hole, as shown in Figure 7d, the following feedback occurs. The smaller the distance between the ball light shell and an adjacent lateral internal region of the hole, the more this area heats up and the greater the force of repulsion of the ball light shell from the side of the internal surface of the hole. As a result, the deformed ball light is located in the middle of the hole, as shown in Figure 6e. Subsequent phases of penetration of the ball light into the room through the hole are shown in Figure 7f, 7g, 7h.

Thus, the OIF, resulting from the initial inhomogeneity of the air temperature in the gap, "pulls" the ball light through a hole in the wall. Ball light recovers its original spherical shape after it has passed the hole and hit the uniform indoor air. Since the thickness of the ball light shell can be extremely thin, the ball light can penetrate through a very narrow gap of the tens of micrometre width.

How Ball Light Catches up Flying Airliners?

The following case that has occurred in the Vologda area in February 1946 is rather indicative. The second pilot has seen how a bright white sphere appeared on the right wing of the plane about the green fire. He has thought that there was a short circuit of an electric lamp, but the flash has not disappeared, as usually happens. The sphere has slowly spread on the frontal edge of the wing and has disappeared under the nose part of the plane. Then a loud crash was heard, and black smoke entered the pilot cabin. Communication was broken. The commander asks the navigator: "Nikolay, can you have noticed where from the sphere was rolled out?" Nicolay said "In fact, it has appeared directly near your legs. I have taken the gun for the rocket to check up, what color a charge within it. But I had no time to open it. The blinding white sphere has flashed at the very same time. It, like an eye of a devil, peered at me, and then has floated to you."

The mix of cases, both new and historical, are collected in the top-secret report of Oak Ridge National Laboratories and presented in the Sagan book [5]. As an example, we present some of them

A large orange fireball hit the windshield's nose. The explosion rocked the ship with such force that the pilots were sure it was a mid-air collision. The ground crew found that the electronics had melted.

The pilots saw "a large fireball flying at great speed." To avoid a collision, the captain fired his airliner upward at a steep angle.

> "Lightning struck a Boeing 757 flying to Dallas/Fort Worth Airport through the co-pilot's window. A bluish-gray fireball the size of a basketball appeared 18 inches in front of the co-pilot's face and hovered motionless for half a minute. Unable to see the dashboard, he bent over to see the dashboard. The luminous sphere slowly and silently disappeared.
>
> It was five past midnight early on the morning of March 18 when a plane flying from New York to Washington was struck by lightning. A bright flash enveloped the entire airliner, and a strong impact woke the sleeping passengers. In front, a scientist from the Electronics Laboratory at the University of Kent in Canterbury, Kent, England, described what happened. The ball passed within a few inches of him and continued on a level trajectory two and a half feet above the floor at a constant speed of three to five feet per second toward the rear of the aircraft. There it disappeared silently. Although the fireball approached a few passengers, none of them felt warm. The ball seemed solid and was in perfect balance."

While flying in a monoplane inside a storm cloud (there was no nearby lightning) we saw distant lightning. Rain began. Glowing green fireballs appeared on the leading edge of the starboard wing close to the aluminum fuselage, then traveled along the front edge to the wing tip then moved straight back in the air.

Unfortunately, the modern science is unable to explain this phenomenon. As quipped Sagan - author of the book entitled "Ball Lightning: Paradox of Physics." All theories have one thing in common - none work.

We have found that all Ball Lightnings that are located closer than 6 meters from the aircraft surface can catch up it. Needless to say, these are very crude estimates that determine only the order of magnitude. However, they show that a flying aircraft attracts all Ball Lightnings located directly in front, as well as on the right, left, above and below in the region of the inhomogeneous atmosphere created by the aircraft.

Reaching the aircraft, the Ball Lightnings stops in the region where the air density is maximal and moves along with this area and, consequently, with the aircraft. No maneuvers of the aircraft can separate the SCL and the aircraft.

The former Deputy Director of the British Meteorological Office was flying during summer to Iraq through dense nimbostratus clouds over the Toulouse Gap at 8,500 feet. A fireball popped through a cockpit window, burned the terrified captain, and glided through the forward cabin. The ball then entered the rear cabin, and while floating down the aisle, it exploded.

It should be noted that the additional air pressure Δp near the leading edge of the wing is proportional to the square of the speed of the aircraft u. Thus, the space from which Ball Lightnings are attracted to the aircraft increases with increasing the aircraft speed. Besides, the probability that the aircraft will gather any Ball Lightnings along its trajectory is proportional to its speed.

As is known, the air is rarefied at great height where aircrafts fly. Because of this an additional pressure is generated in cabins and salons to provide a normal breath for passengers. Since it is very hard and is not justified economically to keep these rooms hermetic absolutely, there is an air turbine compressor that continuously pumps outdoor air into the rooms. It is natural enough to take the air from that region where the air pressure is maximal that is from the region where Ball Lightnings can be located. There is a tunnel between the salon and external space at any position of turbine blades. Since the width of Ball Lightning shell is several micrometers only, a split, the width of which is smaller than ten micrometers, is sufficient for Ball Lightning penetration. The air density gradient is directed into the salon. Moving along the gradient, Ball Lightning penetrates in the salon. Mechanisms that enable Ball Lightning to change its shape are considered in [41]. One can give a recommendation to airplane designers to avoid Ball Lightning penetration. There must be no splits in the regions where additional air pressure is maximal.

In our opinion, the explanation of the mechanisms that allow Ball Lightnings to catch up, accompany aircrafts and penetrate within their rooms is a strong argument in favor of the optical nature of Ball Lightning. Any objects consisting of particles cannot accompany the aircraft. They will blow off immediately by airflow. It does not matter for the light circulating in the region with the highest refractive index whether there is motionless or moving air. The speed of the moving air is negligible small as compared with the speed of light (one million times smaller than the speed of light).

An explanation of this phenomenon is simple and natural. A flying plane creates a large inhomogeneity in the air density around itself. Air pressure near the wing of the plane increases by about 20%. The air density increases accordingly. This inhomogeneity moves with the airliner. The air density increases with approaching the airliner. The circulating light that has got into this inhomogeneity moves in the direction of increasing the air density, that is in the direction of the aircraft and catches up with it.

We should note here two circumstances in the above story. The sphere moved along a frontal edge of a wing, where the air pressure is maximum. The sphere has got into a cabin through the hole into which a gun for a rocket is inserted. This hole is designated to push rockets outdoor and, therefore, the hole is a tunnel between indoor and outdoor spaces. The air pressure in the cabin of the aircraft is greater than that outdoor because air is rarefied at the great height where aircrafts fly. At the same time, the air pressure in the aircruft cabin is maintained at a level close to normal for the comfort of pilots. Therefore, the sphere, moving in the direction of the gradient of the air density, has got into a cabin through a hole into which the gun for the rocket is inserted.

A flying aircraft creates an area of large volume around itself, where the air density increases with approaching towards the aircraft. Once in this zone, the circulating light is directed to the zone of maximum density, that is, to the wing of the aircraft. The greater the speed of the aircraft, the greater the volume of the specified zone and the higher the likelihood of meeting with Ball Lightnings. We can say that a flying aircraft sucks in all the Ball Lightnings that fall into the indicated area, which moves together with the aircraft.

The probability of an aircraft to meet any Ball Lightning in its track is proportional to its speed. Indeed, if the aircraft flies around the Earth in one second, it would gather all the Ball Lightnings encountered during this time. That is why Ball Lightnings are observed in flying aircrafts rather than in aircrafts standing in airports. Are there other hypotheses about the Ball Lightning nature, which could explain how Ball Lightning can catch up a plane and go along with it? We are not aware of such hypotheses.

Why Ball Light Can Move against the Wind

Scientists are even more perplexed by the ability of Ball Lightning to move against the wind. There must be some kind of engine that allows Ball Lightning to overcome the wind resistance. Only creatures and airplanes have this ability. Sagan asked in his book [5] the following. "Ball lightning flying against a breeze or the wind is commonly reported. Why and how? Flying against the wind takes energy. For Ball Lightning without any apparent means of propulsion, this is a mystery and was why scientists doubted Ball Lightning. According to existing physics, there is no explanation at all for how Ball Lightning can fly, let alone against the wind."

Sagan presents the following observations [5]

> "It was 9:30 a.m. during a severe June storm in 1963. We stood at the East window, watching the hailstorm. A severe June storm struck at 9:30 am. It was a frontal storm, with an alleged tornado that turned over trucks

half a mile away, blowing out car windows and damaging our house. The noise was intense. I did not see the fireball materialize, the rain and hail were heavy, and we could see little-just a bright-but-not-dazzling white fireball. I think it would have been dazzling in the absence of the storm. At first, I thought a power line blew down. The fireball was near a cottonwood tree. After the storm, I went out and found no damage. Then I realized that the fireball's drift had been to the south, against the wind. Although I cannot say for certain, it was a sphere; it seemed two feet in diameter. We were 30 feet away from it. It fell to the ground from fifty feet up at a falling speed of 20 miles per hour in a curved path, flying against an intense wind. I have heard eyewitnesses describe Ball Lightning rolling in the Colorado mountains."

This is another observation. The rain was imminent. To the right of a military jeep in convoy, a bright orange ten-inch fireball the color of a sodium lamp appeared between an opening in the wire fence and then floated parallel to the jeep, staying twenty feet from it, about five feet above the ground flying at thirty miles per hour. It glided deliberately despite a fierce wind and vanished after fifteen seconds.

Let us first consider a horizontal movement of the circulating light in the motionless air. If the horizontal velocity of the circulating light equal to zero, the circulating light is motionless because of central symmetry. But this position is not stable. If a small velocity arises, the symmetry is violated because the temperature at the front hemisphere becomes smaller than that at the back one as is shown in Figure 8. It is connected with the fact that the front hemisphere becomes located in the region with a new portion of air, unlike the back one that becomes located in the region where the circulating light heated the air. As a result, hemispheres are located in regions with different temperatures, and optically induced forces applied to them are different.

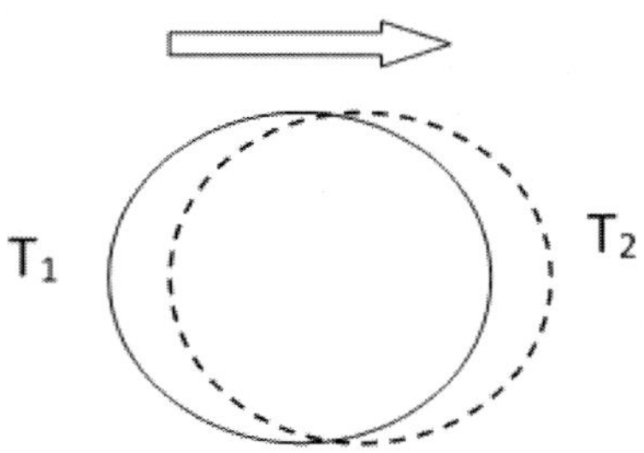

Figure 8. A change of the symmetry in the surrounding air at the shift of circulating light to the right where $T_2 < T_1$.

We can conclude that taking into account the optically induced forces, which are usually neglected because of their apparent weakness, allows us to explain the paradoxes and riddles of natural Ball Lightnings. The inhomogeneity of the air can be caused by the Ball Lightning itself.

Thus, from our reasoning it follows that in still air the Ball Lightning can move in any direction and the force of the air resistance is not a hindrance to it. The greater the energy of the circulating light, the greater its speed. Passing from the coordinate system associated with the air, to the coordinate system associated with the earth, we must add the speed of air, that is, the speed of the wind, to the speed of the ball light. Since the speed of the ball light relative to the air can be arbitrary, the speed of the ball light relative to the earth is also arbitrary. In a particular case, this speed can be directed against the wind. In another particular case, this speed makes any other angle with the wind speed.

The Ball Lightning draws in oncoming cold air, heats it, and rejects the heated air, just as it takes place in direct-flow jet engines. We hope that our explanation of the upwind movement of the ball light is a decisive argument in a recognition of the optical nature of Ball Lightning. A detailed analysis of the movement of the ball light upwind is given in [36].

How Ball Light Penetrates Glass Panes

Currently, it is well established that Ball Lightnings can penetrate through window glass, sometimes destroying them, but in most cases leaving them intact. Damage, if any, is local and is in that region through which Ball Lightnings penetrated. In work [6], 46 cases of penetration Ball Lightnings indoors are described. These are some other observations.

> "During a severe thunderstorm, a sparkling ball 4-5 cm in diameter slowly passed into the room right through the window glass. Passing through the glass, it did not change its shape and passed as if there was no glass. Having hit the metal ball that adorned the headboard, it rebounded elastically towards the window and also slowly left through the glass. When the ball collided with the bed, a melodious sound was heard, reminiscent of the sound of a tuning fork. All this lasted 5-7 seconds. There were no traces in the glass through which the ball passed twice."

> "After a short thunderstorm with heavy rain, a large puddle formed in the yard, and my son and his comrades began to run around it. But at that time a new thunderstorm began to gather, and I called my son home. As soon as he entered the house, I began to wash him legs from a basin standing on the floor. At that time, I heard a crack in the window. I raised my head and saw how Ball Lightning of 8-10 cm diameter was flying through the window glass. Passing through the glass, the ball did not change. It flew straight at us and burst between me and my son (about 15 cm away from me). The sound at the same time was like a shot from an air gun. My son and I remained unharmed. I did not find a hole in the window glass."

Melshina from the Astrakhan region reports a case in which bright blue apple-sized Ball Lightning, in front of dozens of people, went straight through a window pane without damaging it, and struck one of the witnesses. The victim came to his senses only a few hours later.

> "I was sitting on the veranda and putting papers in order. There was a thunderstorm, lightning flashed continuously. Ball Lightning hit the windsock, went in a spiral down the pole to the roof and went under the roof. It was the size of a blue tennis ball and also, it seems, turned around its axis. After a while, a watchman, pale with fear, came running. According to her, Ball Lightning came out of the ceiling and reached the light bulb through a hanging electric wire. It entered the light bulb, came out of it and, falling to the floor, disappeared. The light bulb remained intact, only the glass inside darkened. There were no traces of lightning on the ceiling and on the floor, but a fire broke out in the attic: sawdust caught fire along the path of the lightning."

It should be noted that the vast majority of Ball Lightning theories cannot explain this Ball Lightning property and therefore, must be rejected. This applies to all theories that address any particles: electrons, ions, clusters, plasma, and the like. It is known that particles cannot penetrate through glass. Glass flasks and test tubes used in the chemistry, as well as glass cathode-ray TV tubes, are a good confirmation of this.

Scientists scratch their heads in perplexity. How can Ball Lightning penetrate glass panes without damaging it? What then does it consist of? We are well aware that no material particles can penetrate through glass. What then emits light?

The window is transparent to light, and light can penetrate through the glass. The windows are designed for light penetration. The circulating light freely penetrates through the glass transparent to it. Compressed air cannot penetrate the glass, but this is not required. Exactly the same air is on the other side of the glass. This air is compressed by the penetrating circulating light.

Stages of penetration of the ball light shell through the window pane are shown in Figure 9.

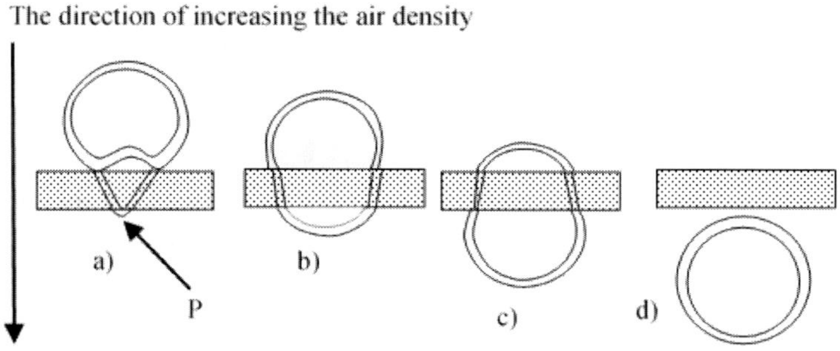

Figure 9. Stages of penetration of the Ball Light shell through the windowpane.
a) initial stage of penetration when the region of the compressed air is produced on the backside of the window pane
b) intermediate stage when a smaller part of the Ball Light shell penetrated on the backside in the room
c) intermediate stage when a greater part of the Ball Light shell penetrated on the backside in the room
d) Ball Light penetrated the room completely.

Let us consider this phenomenon in more detail since it is unknown until now. When Ball Lightning touches the surface of the glass plate as shown in Figure 10, the light propagates along with the plate and the angle of incidence $\alpha = 90^0$. It would seem that the light should also propagate in the room at an angle of 90^0. However, the reflective index of the compressed air in the shell ball light is greater than the reflective index of air on the opposite side of the plate where the normal air pressure takes place. In this case, there is a total internal reflection from the surface of the glass in the room, as is shown in Figure 9. As is seen in Figure 9, the light beam returns into the external air.

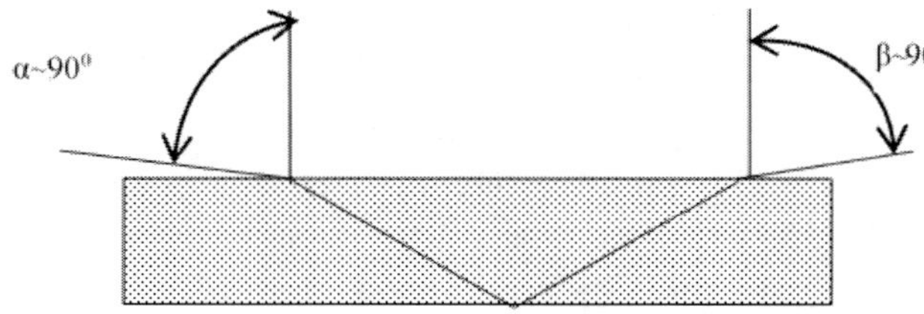

Figure 10. Total inner reflection from the bottom surface of a glass plate.

However, the region where the incident beam enters the plate and the region where the reflected beam leaves the plate are different. A distance between these regions is comparable with the width of the plate. Seemingly, the reflected beam cannot return to the ball light shell. Possibly, a similar situation takes place in nature, and ball light, in this case, does not penetrate through a glass plate. But many attempts are given to the ball light. The situation presented in Figure 9a is possible where ball light heats a local region of the glass due to light radiation and inevitable dissipative losses in the glass because its transparency is not absolute. As a result, the spherical shape of the ball light is deformed. In this case, the light can return into the ball light shell after the total internal reflection from the outer side of the glass as is shown in Figure 9a.

A strong electromagnetic field occurs in the layer of the air near the surface of the glass in the room. Letter *P* in Figure 9a shows the position of this layer. The thickness of this layer is comparable with the average wavelength of white light and is approximately equal to 1 μm. This field cannot excite the light wave propagating in a room due to the phenomenon of total internal reflection. At the same time, this field can compress the air in this layer. The reflective index of the compressed air is greater than that of the uncompressed one. The next portion of the light that is circulating in the ball light shell penetrates the area of the compressed air. This leads to an increase in its volume. But the light cannot propagate in the room because the air density increases with approaching towards the plate. As a result, the light bends in this direction and returns to the ball light shell.

As the volume of the compressed air increases, the area with increased air density turns into part of a spherical film, as shown in Figure 9b. In this case, a part of the circulating light begins to circulate in the room.

If the air densities are different on opposite sides of the glass pane, ball light is moving in the direction where the air density is greater in accordance with the above-described effect. It seems that the Ball Lightning easily passes through the glass, like the light that enters perpendicular to the windowpane. In fact, only light passes through the glass. After that, it forms a layer of compressed air on the opposite side of the glass from the air in the room. A sequence of steps of Ball Lightning penetration through a glass pane is shown in Figure 9.

As is seen, the light within the ball light shell is circulating through both the outer and internal sides of the windowpane. But the compressed air in the ball light shell located on the outer side of the window pane does not penetrate in the ball light shell located on the internal side of the glass. Light cannot be odorous. Therefore, any odors cannot penetrate through glass. An odor after Ball Lightning disappearance may refer to the smell that arose when ball light heated any objects inside the room.

Some evidence indicates that Ball Lightning before entering the room, repeatedly bounces off the glass. This is explained by the fact that the glass is heated due to radiation. In this case, Ball Lightning bounces off the glass as from a conventional opaque obstacle. There is evidence that if the glass is chipped off or there is a gap in the window, then Ball Lightning comes through this gap. In this case, ball light changes its shape in the same manner as ball light penetrates through the usual hole. The number of reported cases, when Ball Lightning penetrates through the crack in the window, is compared with the number of cases where Ball Lightning penetrates directly through the glass. As the number of windows with defects is significantly smaller than the number of windows with conventional panes, it can be concluded that the presence of defects increases the probability of Ball Lightning penetration through a window significantly.

As for the heating of glass during the passage of Ball Lightning, the result depends on many parameters such as the intensity of the circulating light, the losses in the glass, the presence of dirt on the glass, the thickness of the glass plate, etc. Having analyzed the features of the resulting holes in the glass, we can conclude that the glass is heated rapidly enough, and the heating is localized in a circle, through which Ball Lightning penetrates. Assuming that the rate of passage through the pane is constant, we can see that the maximum heating occurs at the circle whose diameter equals tBall Lightning diameter.

What Are the Sources of White Light Radiated by Ball Light?

Indeed, there are no excited atoms in the circulating light that emit photons. The nature of the white light emitted by the circulating light is different from the nature of the white light emitted by a body heated to a high temperature. The white light emitted by the circulating light is its part. Light is emitted due to the phenomenon of molecular light scattering. This is the phenomenon responsible for the blue color of the sky and the orange sun at sunset.

Why Ball Light Disappears Unexpectedly, Instantly and Traceless

What then does Ball Lightning consist of? Perhaps it just becomes invisible or is hiding somewhere?

Two words are enough to explain this phenomenon. This is the "occurrence of instability" [37]. Like a soap bubble that bursts when the film thickness becomes less than a certain threshold, the circulating light disappears when the light intensity, which gradually decreases due to the light radiation, becomes smaller than a certain threshold. Light scatters in all directions. The compressed air expands, and there are no traces left in the place where the circulating light was located.

The same situation occurs with an unexpected avalanche in the mountains, with an icicle falling from the roof. Perhaps a more illustrative example is the loss of stability of a rotating top. Due to the inevitable loss of energy, the speed of rotation of the top gradually decreases. When the rotation speed decreases below a certain threshold, the rotation becomes unstable, a precession occurs, which rapidly increases, which leads to the fall of the top.

Ball light disappears when it becomes unstable (although the stability of the ball light is highly conditional, because of the constant reduction of the stored energy in the ball light excludes any steady-state).

Ball Lightning continuously radiates a part of the circulating light, so the energy stored in it will gradually decrease. The light is radiated due to two types of losses. One type of loss is associated with the molecular scattering of light in air due to the inhomogeneity of the air density that arises due to the Brownian motion of molecules. This type of loss was discovered by Rayleigh, who showed that components of the white sun light in the short-wavelength part of the spectrum scatter more strongly than in the long-wavelength part of the spectrum. This type of scattering explains the blue color of the sky, since the blue components in the composition of white sunlight scatter more strongly than the red ones. The orange sun can be seen when it is low on the horizon. In this case, the sun's rays pass through a thick layer of the atmosphere. The blue rays turned out to be scattered, and the red rays remained in the composition of the transmitted sunlight.

Another type of loss occurs in optical waveguides, which confines the propagation of light in surrounding space and, as their name implies, guide the light in the right direction.

This type of loss is called the radiation loss. For example, optical fibers used in telecommunications are made in such a way that this type of loss is negligible, although it theoretically exists. Radiation losses are determined by two parameters. This is the degree of curvature of the light guide and its refractive index. Radiation losses increase as the curvature increases and as the refractive index decreases. For example, if red light propagates through a fiber, then this light is invisible, that is, it is impossible to determine whether light propagates in the fiber or not. However, if the fiber in which the light propagates is wound around a finger, the fiber becomes red. In this case, part of the propagating light is radiated into the surrounding space due to radiation losses.

As for the radiation of the light circulating in the shell of Ball Lightning, then practically during the entire time of observation of the Ball Lightning, it radiates light due to losses of the first type, that is, due to the molecular light scattering.

As the energy of the light in Ball Lightning gradually decreases, the air pressure produced by the circulating light decreases. As the air pressure decreases, its density and refractive index decrease. In this case, the radiation losses increase, but they continue to be significantly less than the losses due to the molecular light scattering.

However, there comes a point when the radiation loss becomes greater than the loss due to the molecular light scattering. In this case, a positive feedback comes into play. An increase in the radiation losses leads to a faster decrease in the refractive index. A decrease in the refractive index leads to a more rapid increase in the radiation losses. The speed of these transient processes is related to the speed of light. Therefore, the considered processes occur almost instantaneously on a time scale measured in seconds. Observers note that Ball Lightnings do not disappear gradually, but immediately become invisible.

Based on the above, a change in time of the energy stored in the spherical shell can be represented as is shown in Figure 11.

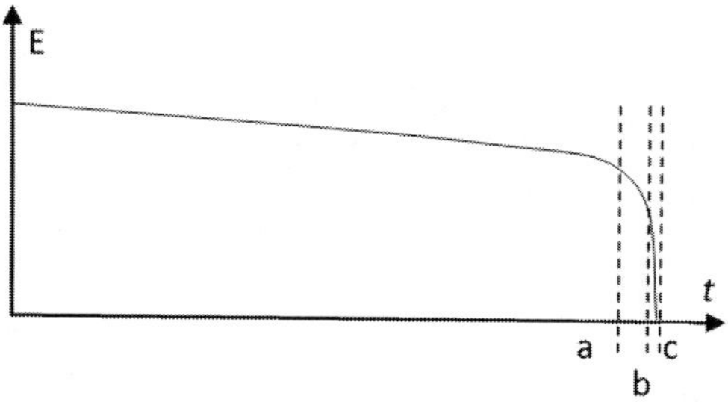

Figure 11. Dependence of the energy E stored in the light bubble on time.

In the time interval from 0 till point *a*, the energy stored in the light bubble decreases gradually due to the Rayleigh scattering in accordance with Figure 11. The radiation losses increase in this time but leave significantly smaller as compared with losses due to the molecular light scattering. In the time interval from point *a* till point *b* the radiation losses become comparable with the molecular light scattering and continue to increase. In the time interval from point *b* till point *c* the radiation losses increase abruptly due to the positive feedback.

As for the suddenness of a disappearance of Ball Lightning in any part of its trajectory, it is impossible for an observer to determine the moment when the instability of the circulating light occurs. Such instability occurs when the energy density of the circulating light becomes less than a certain threshold. This threshold differs little from the stationary value at which the circulating light is stable. Therefore, the observer cannot determine the difference between the threshold and the steady state. He sees that Ball Lightning shone without changing its brightness, and then disappeared for no apparent reason.

A similar situation occurs when a soap bubble disappears. The thickness of the soap film gradually decreases. In this case, the volume of the soap bubble begins to increase due to the excess air pressure inside the film, which leads to an increase in the surface of the film and a decrease in its thickness. In this case, in the final analysis, a moment comes when a positive feedback comes into play, at which a decrease in the film thickness is accompanied by an increase in its surface, which contributes to a further decrease in the thickness. Eventually the film breaks and the bubble bursts. The observer also cannot determine the moment when the bubble will burst.

Quite often it is noted that a disappearance of Ball Lightning is accompanied by a sound in the form of an explosion. At the same time, many witnesses specify that the sound resembles a shot from a pistol. It can be assumed that the specificity of the sound of a pistol shot is due to the fact that a barrel of the pistol is relatively short. Therefore, the duration of the sound when a pistol is fired is relatively short and is associated with the expansion of a relatively small volume of air. A similar situation takes place during an expansion of highly compressed air in a spherical shell of Ball Lightning. Due to its small thickness, its volume is also small.

Why Ball Light Can Bounce

Bounces of a conventional rubber ball on a floor occur for two reasons. This is the mass of the ball that is attracted to the floor by gravity. This is an elastic collision of the ball with the floor, in which the ball first deforms, in which the force acts on the ball, opposite to its downward movement. As a result, the speed of the ball gradually decreases to zero. Then the same force accelerates the ball up. As the ball moves upward, the speed of the ball gradually decreases as its kinetic energy is converted into potential energy. Eventually the ball's speed becomes zero, the ball stops and starts moving down. A similar situation occurs with a stone thrown vertically upward. The process is repeated. To carry out this process, it is necessary that the ball has a mass and is in the Earth's gravitational field.

The mass of the ball light also exists and is equal to the mass of compressed air in the shell. The ball light is in the field of gravity and gravity also acts on it. In addition, the ball light is located in an inhomogeneous optical medium in the form of air, the temperature of which is minimal near the floor. The ball light is affected by an optically induced force directed towards the floor. This force exceeds the force of gravity, so the ball light approaches the floor much faster than a balloon filled with air.

Generally speaking, the ball light shell is elastic, since it restores its shape in a homogeneous optical medium. But this elasticity is small enough to provide the necessary upward acceleration when the ball rebounds. The rebound of the ball occurs for other reasons. When the ball light moves towards the floor, the floor is considered as an obstacle. At a short distance from the floor surface, the ball light heats this surface due to its light radiation. In turn, the surface heats the layers of air close to it due to the phenomenon of heat conduction. As a result, the air temperature in the gap between the ball light and the floor surface increases significantly. The air temperature is maximum at the floor surface. Therefore, an optically induced force arises, directed to the region where the temperature is lower, i.e., directed upwards.

This force is analogous to the elastic force that pushes the rubber ball off the floor. As a result of the action of this force, the speed of ball light changes to the opposite one and ball light begins to move upward. The vertical speed of the ball light gradually decreases due to the force of gravity. The vertical speed at a certain height becomes zero, and the ball light begins to move downward. The ball light is ready to repeat the next jump.

There is one refinement to the considered process, which can be understood if we take into account the processes that occur when the ball light moves against the wind. This refinement consists in the fact that the air resistance during this movement is compensated partially by the action of the optically induced force arising because the air pressures on the front and rear hemispheres are different. As a result, the ball light jumps to the same height with which it fell. This phenomenon is considered in more detail in Chapter 2 when considering the jumps of other objects from the world of circulating light.

Why Ball Light Seems Cold

One additional argument in favor of the optical nature of Ball Lightning is an explanation of the following apparent contradiction. Most physicists, if not all, wonder why Ball Lightning emits white light, the spectrum of which corresponds to the temperature of the hot glowing body of several thousand degrees, and at the same time Ball Lightning is relatively cold, as it does not burn the objects around it. This paradox has a simple explanation.

The light that emits the ball light is not the light that radiates a bright hot body. The light radiated by the ball light is the light arising due to the molecular light scattering of the intense light circulating in the ball light shell. Indeed, the bright light was generated by excited atoms of the air heated to a high temperature during the gas discharge of the linear lightning in which Ball Lightning appeared. This light fell into the ball light and has existed for a relatively long time. Thus, the nature of the light emitted by the ball light is different from the nature of the light emitted by a bright hot body. The ball light has caught and preserved the light radiation of an ordinary linear lightning strike and gradually emits it in a form of scattered light.

Why Ball Light Moves along Walls

Sagan devotes an entire section describing a motion of Ball Lightning on a perimeter of the room [5]. There are some of these cases.

Lightning struck a tree several hundred feet away, and a six-inch ball entered an open end of a garage, followed the wall counterclockwise around the garage on a ground level. It faded as it left the garage.

An orange ball flew around a radio transmitter room. Lightning struck the radio transmitter antenna. I was in the transmitter room. The transmission equipment was destroyed and melted. Ball Lightning was not a flash, but an orange sphere of flame that quickly moved across the room, leaving behind the smell of ozone.

After an earlier storm in early June, in the weather hot and humid, just before sunset, a man walked into his garage. A basketball-sized blue fireball with a large electrical coating floated on the left side, floated right, and then back to the left like a caged animal, then vanished. The garage was dark, and the doors were closed. Was the ball there before he opened the door? Was the ball trying to escape?

Stakhanov in the case 29 describes a movement of Ball Lightning along a closed path in the trailer where the builders lived [2]. The Ball Lightning trajectory in the trailer is shown in Figure 12. As expected, the Ball Lightning, approaching the next wall, changes its direction of movement and starts moving again along the other wall.

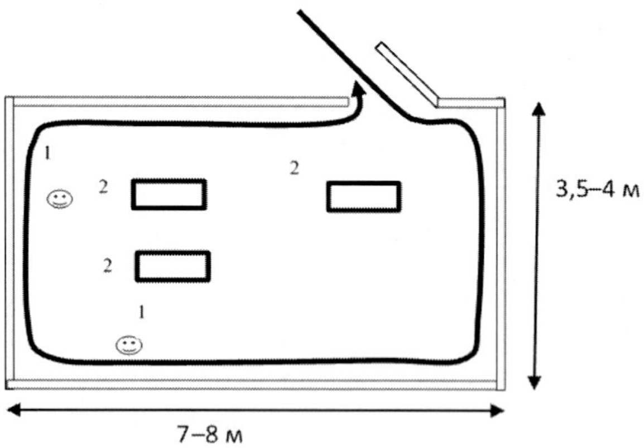

Figure 12. Trace of the circulating light motion in the shed of builders. 1 – observers; 2 – tables.

Usually, the walls are colder than the air in the center of the room because the walls are connected to the ground, which is colder than the surrounding air. If the room has a stove, it first heats the air around it, and then the air heats the wall. Cold walls are an obstacle to the rectilinear movement of the ball light. When trying to bypass the obstacle, the ball light changes its direction of movement and tries to bypass the next wall.

Why Ball Light Moves in a Room Near a Floor

It is easy to explain the features of the Ball Lightning motion indoors. As is known, the air inside a room is closer to a steady-state than outside. Since warm air is lighter than cold air, the air near the floor is cooler than the air near the ceiling. Thus, has penetrated the room, the ball light moves toward the floor, which in this case acts as an obstacle. There are many reports that the Ball Lightning, penetrated the room, moving around the floor. In some cases, the motion is accompanied by bouncing. The explanation is given above.

Why Ball Light Moves near Metal Objects

There are numerous reports that Ball Lightning is attracted to a metal object, such as a fence of metal bars or telephone wires. Approaching a metal object, Ball Lightning usually starts moving along the object. Let us try to explain this feature of Ball Lightning, bearing in mind that the ball light is moving from hot air toward cold air.

It is known that metals are not only good conductors of electric current, but also are good heat conductors. The temperature of various objects on the Earth's surface changes due to daily fluctuations in air temperature. On a sunny day, the temperature of the upper layers of the Earth's surface is higher than that of the deeper layers. For example, if a metal column is embedded in the ground, the temperature of the column is lower than that of the ambient air because the heat that the column receives from the ambient air goes down into the ground. As a result, the air temperature near the column is lower than outside. Ball Lightning, appearing near the column, begins to approach the column. However, it cannot touch the column, since at a small distance the air between it and the column heats up by ball light. At the same time, it cannot move away from the column, since the air near the column is colder than at a great distance from it. As a result, the ball light is located at some distance from the column. It can neither approach nor move away from the column since forces arise that return it to its original position. A similar situation was considered above when the ball light moves at a certain distance from the surface of the Earth.

For example, Stakhanov [2] in case describes the observation, in which the two Ball Lightning jumped out of the oven 5 (Figure 13), passed through two doors, passed along two walls, and pulled to ground 6 in the form of a metal wire. Their trajectory is shown in Figure 13. The motion of Ball Lightning is observed by four observers: 1, 2, 3, 4. Approaching the metal wire 6 near windows 7, Ball Lightning changed its shape, becoming a prelate ellipsoid, penetrated through the gap between the wire and the floor, and disappeared. That is exactly what should be happening if keep in mind the above-mentioned features of the moving ball light indoors and near metal objects. Ground 6 in the form of a metal wire is the coldest object in the room.

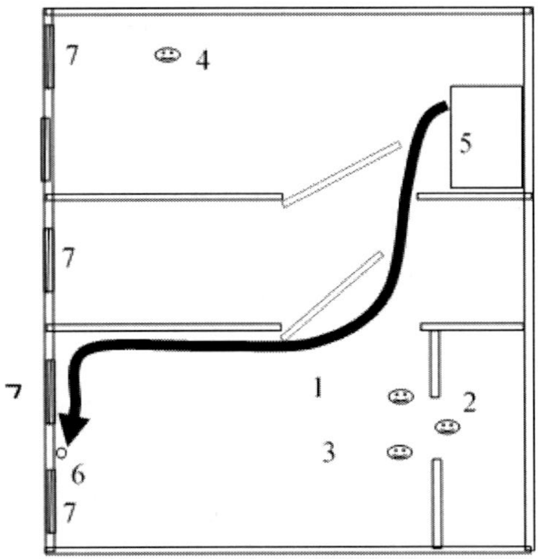

Figure 13. Trace of circulating light from the oven to the grounded metallic rod.

How Ball Light Burns out Craters in Metal Objects

There is a lot of evidence that Ball Lightning partially evaporates the metal in contact with it. For example, a very strong lightning strike destroyed the copper antenna shortly after it was installed for the study of thunderstorm electricity. The witness, who watched the storm, saw a big ball of fire, apparently formed due to vaporizing 65 m two-millimeter wire.

Stakhanov, in case 4, describes a case where Ball Lightning burned through a hole 5 mm in diameter and 3 mm in depth at the end of a cleaning rod attached to a karabiner. It is noted that no traces of melting were observed. The metal just disappeared. This phenomenon can be explained by considering the following two circumstances. First, the refractive index of metal vapors is greater than the refractive index of air. Secondly, objects elongated in one direction can be heated much more than objects, the dimensions of which are approximately the same in all three dimensions. It is most simple to heat a long rod from its face since in this case heat is removed only in one direction along the rod axis. Moving along the rod and approaching its face to a minimum distance, the ball light can heat the face of the rod to the temperature at which the metal began to evaporate. The refractive index in the metal vapor region is maximum. Ball light is located in this area until its energy runs out. In some cases, ball light leaves just a deepening in the rod. In other cases, ball light vaporizes tens of meters of wire, which can be viewed as a very long rod.

In the description of aircraft collisions with the Ball Lightnings is often reported that during the examination of the aircraft, several damaged or melted rivets in front of the fuselage are revealed [5]. This is easily explained if we bear in mind that the rivet located on the smooth surface of the fuselage breaks uniform airflow over the fuselage. Excess pressure is raised near the rivet, and there is a local maximum of the air density. Finding itself in this maximum, Ball Lightning vaporizes a portion of the rivet. Thus, the same mechanisms leading to the evaporation of a long metal cylinder take place. Evaporation takes place until the protruding portion of the rivet disappears. This leads to the disappearance of the excess pressure produced by the protruding portion of the rivet.

Singer, in his book [1], shows the results of the interaction of Ball Lightning with the propeller blades (Figure 14). Because the propeller pulls the airplane forward, the propeller is undergone overpressure. The pressure is maximal in that part of the blade, where the violation of the blade edge occurred. Located in the area with maximum pressure, Ball Lightning rotates with the blade and burns the craters on it, just as it burns craters on fixed metal objects. We can say that Ball Lightning indicated to the developers of the propellers the place where the air pressure on the propeller is maximum.

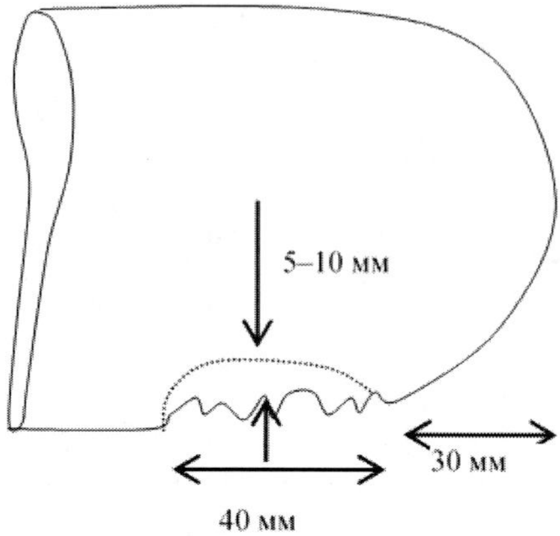

Figure 14. Damages of propeller blades.

Upon closer examination of these cases, the following feature was noted. Evaporation of metal takes place in metal objects, the dimensions of which in one direction are much greater than the dimensions in two other directions (metal wire, copper core in the cable, rivets on the fuselage of the aircraft, cleaning rod attached to a karabiner).

What are the specifics of these objects? How do they differ from other ordinary metal objects? The difference is that their shape provides heating to a greater extent, since the presence of only one dimension of the three possible provides a greater heating of the object. Heat can only travel in one dimension. Therefore, the cooling of the heated area due to the phenomenon of heat conduction occurs to a lesser extent, and therefore the heated area can be heated to a higher temperature.

As will be shown in Chapter 2, metal vapors exist near the surface of a metal object. They increase the reflection coefficient in the space where they are located. Therefore, if the circulating light begins to evaporate the metal, then is located in the region with the maximum reflective index and cannot move from there, since the reflective index in this case will decrease. Evaporation occurs until the energy of the circulating light is exhausted.

The effect of the interaction of the ball light with metal objects is noticeable in the case when the ball light can heat the area of the object to a temperature at which the evaporation of the metal begins. Therefore, it is not surprising that the ball light leaves traces of fusion on the protruding parts of metal objects.

The following case in this respect is very revealing. Ball Lightning burnt out the TV cable with copper core and polyethene shell at a distance of several meters, leaving the shell almost intact.

This fact is easily explained. A scheme of the cable is shown in Figure 15. The metal inside the shell is evaporated so quickly that the area with high temperature is moved along the cable at high speed. This heats only a small area of the inner shell, directly adjacent to the metal core. The heat did not have time to spread to the entire mass of the shell. The highest pressure and thus highest reflective index are near the end of the evaporated wire because the reflective index of metal vapors is maximum. Therefore, Ball Lightning has a form of a thin sausage located near the end of the wire during the whole process of evaporation, as is shown in Figure 15.

The witness saw the appearance of a rolled bright mass of 1 m in diameter, and behind it, a few smaller balls after extremely strong lightning strikes a telephone pillar on the street. Later it was discovered that a wire between two pillars interrupted and a significant chunk lacks [2].

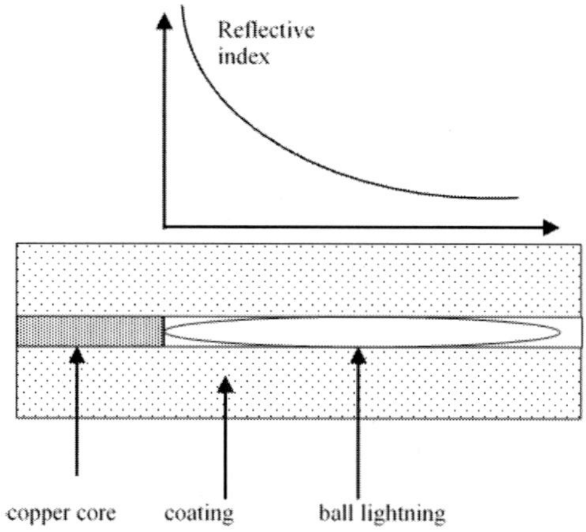

Figure 15. Burning copper core of TV cable by Ball Lightning. Dependence of the reflective index on length is shown in the upper graph.

Why Ball Light Hisses and Causes Radio Interferences

Being at a steady-state, a ball light shape is unchanged, and there is no noise from the ball light. Considering the ball light as an elastic shell, we need to recognize that there is a majority of natural frequencies for mechanical oscillations of the shell. These oscillations may be excited by various inhomogeneities which the ball light meets at its way, in particular, particles of dust. If the natural frequencies are unevenly distributed along the sound spectrum a hissing sound may be heard. A similar sound may be heard in a radio dynamic if the gain of an amplifier is greater than the nominal one.

As for radio interference, many eyewitnesses have noted the effect of ball light on a radio receiver. But this does not mean that ball light generates radio waves, as many scientists believe. The crackling sound from the ball light is also heard in the telephone tube, which, as you know, is not able to receive radio waves. The following explanation can be offered. The ball light can be electrically charged. Mechanical vibrations of its shape at sound frequencies cause a redistribution of the electrical field in the surrounding space. This redistribution is also responsible for the electrical effects in explosions of Ball Lightnings. The same effect causes the crackling sound in a radio receiver and a telephone tube. You can recall the crackling sound in a radio receiver when an electric lamp is turned on or off.

Why the Ball Light May Be Variously Colored

There are many reports in which ball light colors are various. Along with this, there are many various explanations of these observations. Some believe that the significant variability in the color of Ball Lightnings is due to trace impurities.

We give our explanation why ball light color can be various beginning from reddish and ending bluish whereas the color of usual linear lightning is white. Most simply to answer this question, it would be possible so. The temperature of the area in which a ball light appearance takes place can be the most various depending on conditions of gas discharge. As is known, the maximum of a spectrum moves in a short-wave part of a spectrum at an increase in temperature. The ball light represents a combination of the compressed gas and intensive electromagnetic radiation. A maximum of the spectrum of electromagnetic radiation can change over a wide range. The ball light color, which is defined by a maximum of this spectrum, can be too over a wide range. On this simple explanation, various subtleties can be imposed. For example, ball light appears as a result of numerous merges of tiny ball light of small diameter which radiating losses depend on the wavelength of the light circulating in them. Most quickly, such ball light loses the radiation in a long-wave part of a spectrum. As a result, the average wavelength of the spectrum moves in a short-wave part. In this case ball light, formed because of merging numerous tiny ball lights, has a violet shade. Such ball lights were observed in experiments [98].

Besides a steady-state between energies of light and compressed air at the moment of ball light generation is absent. If the initial air energy is greater than that at steady-state, then a part of the air energy is transmitted to light, and its energy increases. Since the total number of photons in light is unchanged in this process, the energy of each photon increases, and, therefore, the light frequency increases also. In this case, the light wavelength is shifted on the blue side of the spectrum. If the initial air energy is smaller than that at a steady-state, the light wavelength is shifted on the red side of the spectrum.

Why Do the Smells of Ozone, Nitrogen Oxides, Sulfur Appear after the Disappearance of Ball Lightning?

Our initial notion that a shell of the circulating light is composed of air is idealized to simplify the original explanation of the Ball Lightning nature. Of course, other gases can also be present in the shell. The question is which gases are preferred. The answer is those gases whose reflective index is greater than the air reflective index. The circulating light tends to increase the reflective index of the medium where it circulates. One way is to increase the reflective index and density of the air by means of compression. Another way is drawing into the shell gases, the reflective indexes of which are greater than that of the air. It will be shown below that the reflective indexes of all mentioned gases are greater than that of the air.

Besides, it has been shown that the energy required to increase the reflective index in the circulating light by by means of drawing is much smaller than the energy required to increase the reflective index by means of the air compresssuion.

It will be shown later that ozone, nitrogen oxides are produced at gas discharges at the moment of arising Ball Lightning, in particular, at corona discharges. A smell of sulfur is the smell of sulfur oxides that are produced at erosive gas discharge. A case is described when a luminous ball was formed by a bolt of linear lightning struck a tree. The ball began to approach the surface of the Earth. After the disappearance of the ball, a hot yellow mass left.

The fact that some gases have the reflective index greater than that of the air affects the behavior of circulating light in air containing such gases. Most often, there is an excess amount of carbon dioxide in the air, which occurs during the combustion of hydrocarbons. The reflective index of carbon dioxide is 1.000412, while the reflective index of air is 1.00027. Previously, we assumed that the inhomogeneity of the reflective index near the Earth's surface is mainly due to the inhomogeneity of the air temperature. Now we must clarify this assumption and take into account the excess content of gases with an increased reflective index in the air, in particular, carbon dioxide. Calculations show that an increase in carbon dioxide content by 1 percent leads to the same increase in the reflective index as a decrease in the air temperature by about two degrees.

How to Protect Yourself from Ball Lightning

Taking into account the above consideration, we can draw certain recommendations on how to avoid damages from Ball Lightnings. The most radical way is to decrease air density in the room. In this case, the outdoor air is drawn in the room, but the ball lights, contrary to expectations, are pushed off the room. An increase in room temperature results in the same effect. If the ball light appears in a room that it is not recommended to wave your hands or, for example, to drive it away by a broom because a region with the excess air pressure is generated between the ball light and the broom. The ball light will approach the broom. It is not recommended to blow in the ball light because the ball light will travel to the region with the excess pressure. This region is a source of wind.

One can move a plane non-transparent object closer to the ball light, for example, a great book with a dark cover or any object like a pillow or a blanket. After heating the air layer between the ball light and the object, the object will push off the ball light.

If Ball Lightning is encountered in a field, then it is necessary to lie on the ground or climb a tree, since Ball Lightning moves at the height of several meters.

Unknown Properties of Ball Lightning Derived from Properties of Circulating Light

We have shown that anomalous properties of the natural Ball Lightnings observed by eyewitnesses can be explained on the assumption that Ball

Lightning is the ball light. Having the ability to predict a behavior of the ball light in a variety of situations, we can predict a behavior of Ball Lightnings in several situations that have not yet been observed.

Ball Lightning of Large Diameter Takes the Form of a Flying Saucer

Consider what shape acquires the ball light of several meters in diameter. At a steady-state, the ball light is located in the air layer, where the air density is maximal. The upper and lower poles are located in layers where the air density is smaller. As a result, the optically induced forces applied to the upper and lower poles are directed to the layer with maximal density, that is to the equator (Figure 16). Thus, there are forces that tend to squeeze the ball light. The forces that provide a spherical form of the ball light are relatively small and decrease as the diameter increases. At the same time, the forces that tend to squeeze the ball light, increase with increasing its diameter. Indeed, the larger the ball light diameter, the greater its tendency to be squeezed. As a result, the ball light takes a shape of an oblate spheroid. If the distributions of the reflective index above and below the equator are different, the flattening of the top and bottom of the Ball Lightning are also different. As a result, the ball light takes the form of a body of revolution, which resembles a saucer or a plate.

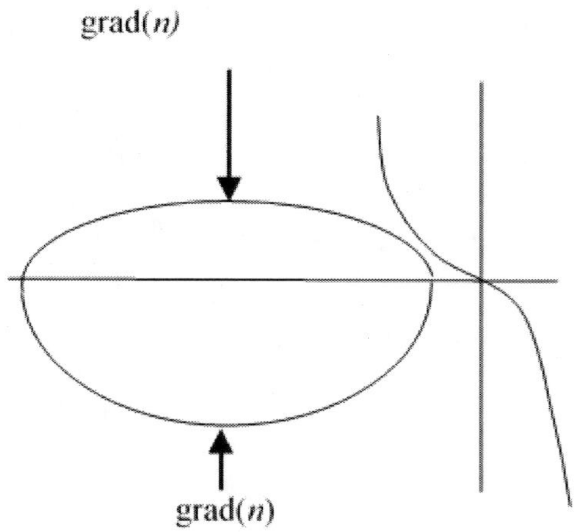

Figure 16. A form of the ball light of great diameter located in the air layer where the air reflective index has a maximum. Dependence of the gradient of the reflective index on the height is shown at the right hand.

It is known that unknown flying objects (UFO) can move at great speed and change a direction of their speed so sharply that the overload arising in this case that is not permitted for living beings.

UFOs move very quietly or silently, unlike jet aircraft. They may not produce a sonic boom during supersonic motion. UFOs exhibit linear accelerations within one second to speeds exceeding those of conventional aircraft. They are also capable of maneuvering at sharp angles.

For example, during the 2004 Tick Tok incident, a UFO moved (on radar) from 80,000 feet to various altitudes from 28,000 feet to 50 feet above sea level in 0.78 seconds. This gives an acceleration of more than 12,000 g (g=9.8 m/s^2 is the acceleration of free fall) and a speed in excess of 100,000 miles per hour, assuming that half the distance was used for acceleration and the other half for deceleration. If UFO is the ball light, it can move with great speed, provided that the reflective index in the atmosphere is changed sufficiently quickly (in particular, UFO can catch up on a flying aircraft). If the atmosphere is not stationary, the UFO may encounter an area in which the direction of a change in the air density changes quite abruptly. In such places, the UFO sharply changes a direction of its movement [1].

Ball Lightning Motion in Vortices, Tornadoes, Cyclones

Ball light moves along the gradient of the reflective index which direction can be different from the direction of the airspeed. A particular case of vertex a movement of air is presented in Figure 17. In this case, the velocity of the air is directed on a tangent to the circle. As to the air pressure, it is identical on this circle owning an axial symmetry of the system. Then the tangential component of the gradient of the pressure is equal to zero, and the gradient of the pressure is directed along the radius. The air pressure decreases at an approach to the center. As the air density is proportional to the air pressure with other things being equal, the air density increases with increasing the radius and the ball light, in this case, moves along the radius from the center perpendicularly to the wind. As a result, Ball Lightning is pushed out of the vortex.

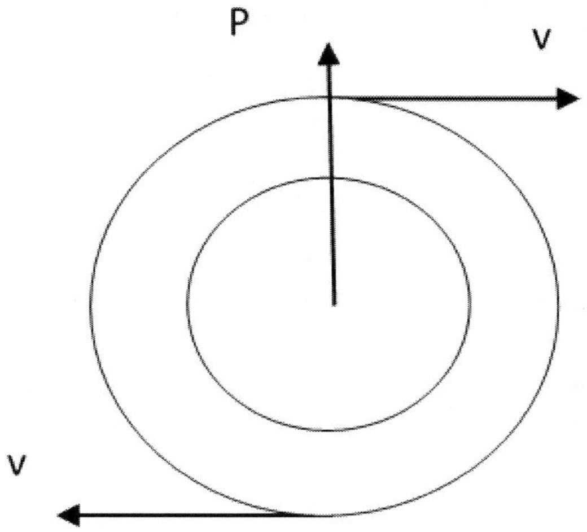

Figure 17. Directions of the airspeed and the gradient of the air pressure in vertex.

Ball Lightning Motion in Gas Pipelines and Steam Pipelines

As is known, if the air moves in some pipe, the air pressure increases in the direction that is opposite to the air velocity. In this case, unlike amovement of all known objects along with the air velocity, the circulating light moves in a pipeline against the air motion. Indeed, the air moves from the space with the high pressure to space with the lower pressure. On the contrary, the circulating light moves from the space with the low pressure to the space with the high pressure. Once in such a pipeline, the ball light will move in the direction opposite to the gas movement. Of course, such a movement is impossible to observe. However, there is a reason to believe that such a movement took place during the Chernobyl accident.

Six seemingly unrelated and inexplicable random facts are connected with the Chernobyl accident. Just as the famous detective Sherlock Holmes was able to determine the pattern and causes of the crime and point to criminals using minor details that others did not take into account, using his deductive method, we, taking into account the properties of the ball light were able to connect these random factors together, identify the cause of the explosion and explain some of the events that took place one minute before the explosion.

Physical Conditions Responsible for the Existence and Anomalous Behavior of Ball Lightning

We remember our promise not to give a single formula. It is not so easy to explain a mysterious phenomenon that has not been explained for many centuries. On the other hand, the nature of Ball Lightning is very simple and obeys well-known physical laws known as far back as the 19th century. Therefore, our explanation, we hope, will be clear to a wide range of readers. The reaction of any sane person to our explanation that Ball Lightning is a bubble made of a film of highly compressed air, in which ordinary light rotates in all possible directions and compresses the air in the film, is as follows. Why should such an object be stable? Compressed air tends to expand and circulating light tends to expand and travel in a straight line. All this formation will expand endlessly and disappear without a trace.

The more trained reader will present his objections in a more scientific manner. It is known that the total energy of any system is minimal in any stable system. In the case under consideration, the energy of the compressed gas decreases as it expands, and the energy of the circulating light also decreases. How can a minimum of energy be formed in this case? For a minimum, it is necessary that at least one of these energies increase with expansion. This is a necessary condition. But it does not work in this case.

The reader can ask. Are there some laws unknown to us that ensure the existence of such objects? If yes, which ones? If not, why is such an object stable?

We can answer. No new laws are required to explain the stability of this phenomenon. Knowledge of quantum mechanics, nuclear physics, thermodynamics, etc. is not required. Enough knowledge known in the 19th century.

You just need to know how to apply them. Indeed, when light circulates in a film of compressed air, it produces centrifugal pressure on the film. When the film expands, the light produces positive work and, therefore, its energy decreases. However, light does not circulate in free space, but it is circulating in an optical medium in the form of compressed air. As the film expands, the air pressure drops and the refractive index of the optical medium drops also.

The energy of the circulating light depends om the reflective index of the optical medium where it is circulating. With a decrease in the refractive index, the energy of the circulating light increases. Therefore, under certain conditions, the energy of the light circulating in the optical medium can increase.

We determined these conditions. It turned out that stability occurs at an air compression of 1000 or more. Compression is understood as a factor that determines the degree of reduction in the air volume relative to the volume the air occupied under normal conditions. When air is compressed 1000 times, its pressure increases not 1000 times, but 20,000 times, that is, it reaches 2 GPa.

This conclusion about the very great air pressure is confirmed by the explanation of another feature of Ball Lightning. The lifetime of light in Ball Lightning is more than a million times longer than the lifetime of light in miniature glass balls used in optics as optical resonators of high quality factor. Analyzing the scattering of light on the basis of known laws, we came to the same conclusion that such a lifetime of light in compressed air is possible only if the air is compressed to a pressure of several GPa [7, 51].

From an analysis of the stability conditions for circulating light, it follows that the energy of circulating light is 10–30 times greater than the energy of the compressed air. It should be borne in mind that both the energy of the compressed gas and the energy of the circulating light are concentrated in a very thin spherical film. Therefore, the energy density in a spherical film is extremely great, which allows Ball Lightning to leave craters or burn through metal objects.

The existence and anomalies of circulating light are based on two phenomena discovered in the 19th century.

The first phenomenon discovered by Faraday is the occurrence of forces acting on a dielectric in an electric field. This phenomenon can be observed when pieces of paper (dielectric) are attracted to an electrolyzed comb. It has been shown that the acting force is proportional to the square of the electric field strength.

The second phenomenon discovered by Helmholtz is the compression of a dielectric placed in an electric field. The degree of compression is also proportional to the square of the electric field strength.

Since light is an electromagnetic wave in which there is an alternating electric field, a transparent dielectric (optical medium) through which the light propagates will be subject to these phenomena. Moreover, the greater the energy density of light, the stronger these phenomena are manifested. At ordinary energy densities of light, when light propagates linearly, the manifestations of these phenomena are so weak that significant efforts are required to fix their existence. However, in circulating light, its energy density increases by a factor of billions. The manifestation of these phenomena also increases to the same extent. It can be said that these two known phenomena are beginning to fully manifest themselves in the world of circulating light.

They are responsible for the existence and anomalies of objects in the world of circulating light.

The first phenomenon is responsible for the appearance of optically induced forces, the action of which makes it possible to explain all the anomalies, paradoxes and mysterious behavior of natural Ball Lightning.

The second phenomenon is responsible for the compression of air in a thin spherical layer in which light circulates in all possible directions. This phenomenon is now called optical electrostriction pressure. It is shown that this name does not correspond to the physical nature of the phenomenon that takes place in an electromagnetic wave. It would be more correct to call the phenomenon in the field of circulating light as electromagnetic compression, since not only the electric field of the electromagnetic wave, but also the magnetic field takes part in the compression of the optical medium. However, these are already details that were refined during the development of the theory of Ball Lightning.

Conclusion

Everyone had the opportunity to make sure that the behavior of the circulating light based on the simple law, according to which the circulating light moves in the direction of increasing air density, and the behavior of natural Ball Lightning, based on numerous eyewitness accounts, are completely identical. Everyone must decide whether the circulating light exists in nature. We only note that the knowledge available in the 19th century is quite enough to be convinced of this. We hope that the reader, having read this chapter, will recognize that our notion about the natural Ball Lightning works and the Ball Lightning nature has been discovered. Certainly, it is hard to believe that a circulating light in the form of ball light can exist in nature. Nobody even hinted about such a possibility.

Should recognize that the properties of the ball light and its behavior in the atmosphere can be quite unexpected. The mysterious and inexplicable behavior of natural Ball Lightnings tells us about the possible features of the ball light behavior. Having analyzed these features based on the simplest law of optics, we were pleased to note that the behavior of our ball light completely coincides with the mysterious and intriguing behavior of natural Ball Lightning.

The above century was required for the Copernicus theory to win a general recognition. General sense spokes that the Sun and stars move but the Earth is motionless. Only scientific knowledge was enabled to override this misconception.

From the presented consideration, it is necessary to conclude that the ball light is not a game of the mind, but it is an objective reality. Till now the circulating light was studied neither theoretically, nor experimentally. Anybody did not suspect at all about an opportunity of its existence. Ought to underline that the existence of the circulating light in nature is based not on theoretical rationale but on the wonderful coincidence of behaviors of the ball light and the Ball Lightning in the air atmosphere as is shown in this chapter.

Chapter 2

The World of Circulating Light

In every branch of knowledge, the progress is proportional to the number of facts on which to build, and therefore to the facility of obtaining data.
J. Maxwell

Introduction

While experts shook their heads, moved their lips and expressed doubts about the correctness of the optical nature of Ball Lightning, we had a unique opportunity to study other phenomena from the world of circulating light without fear of competitors. It turned out that there are many other objects in which circulating light plays a major role. Most of these objects are obtained artificially. However, there are good reasons to believe that circulating light (or rather, circulating waves of a nature unknown to us) takes part in the phenomena of the micro world.

Unlike natural Ball Lightning, which is a rare unpredictable phenomenon that cannot be systematically studied experimentally, other objects with circulating light are obtained mainly at attempts to reproduce Ball Lightning in a laboratory. The properties of such objects have been carefully studied and published. Only the study of this information made it possible to discover a number of new phenomena and understand better the properties of the circulating light. These issues are discussed in this chapter.

Thousands of attempts have been made to obtain a Ball Lightning in a laboratory. The most interesting attempts made before 1980 are described in Barry's book [3]. However, neither the researchers nor Barry himself could unambiguously determine whether the resulting objects belong to a family of Ball Lightning. The behavior of many objects could not be explained based on current knowledge. But this does not mean that such objects belong to a family of Ball Lightning.

Having known the properties of the circulating light, we can try to assess whether questionable anomalous objects are associated with circulating light, which includes a wider class of objects, not just Ball Lightning.

What are the signs of circulating light?

First, it is the presence of a clear boundary of the luminous object, which is formed by the shell of circulating light.

Second, unlike the blinding light of the voltaic arc, the circulating light is connected with a dim white glow that can be observed with the naked eye.

Third, the radiation comes from the surface. This property can only be observed for large enough objects.

Fourth, there is no great heat near the object

Fifth, in some cases, luminous objects burn through a foil or leave craters on it. At the same time, they bounce off a plain sheet of paper. This also indicates the presence of circulating light in the object.

In general, any anomalies of luminous objects that are similar to the anomalies of natural Ball Lightning can indicate the presence of circulating light in the object, since all the anomalies of Ball Lightnings are explained by a presence of the circulating light in them.

Attempts to produce Ball Lightning in XIX-XX centuries are described in books of Barry [3] and Singer [1] in detail enough. The solving contribution to the production of artificial Ball Lightnings has been made in the USSR at the end of 80th - the beginning of 90th years. Results of these works in a form of about a hundred correspondences are presented in the book "Ball lightning in a laboratory," published in 1994 [89]. Additional attempts to produce Ball Lightning have been undertaken in the 21st century also.

Researchers were not able to answer the question of whether their luminous objects belong to Ball Lightning. They noted that these objects have properties that they cannot explain with the application of the known laws of mechanics and optics. We will call all these objects abnormal or anomalous glowing objects (AGOs). Since the physical nature of either Ball Lightning or AGO was unknown, it was impossible to judge a relation between Ball Lightning and AGO.

It turns out that the anomalous properties of AGO can be explained on an assumption that there is the light circulating along the AGO surface in all possible directions. A reason for the arising of the circulating light as well as the means that provide the circulation of light and prevent its radiation in free space can be various. But the circulating light demonstrates tts anomalous properties in all known cases. We will try to present the anomalous properties of AGOs and to show that the circulating light is responsible for these anomalies. In other words, we are going to show that anomalous objects belong to the world of circulating light. This gives additional arguments in favor of our optical theory of natural Ball Lightning.

Experiments on the production of artificial Ball Lightning are very valuable because, unlike natural Ball Lightnings, parameters of AGOs produced in the experiments can be carefully measured and AGOs can be repeatedly generated. Various researchers differently call objects studied by them: the power-intensive plasma formation, the luminous independent object, luminous abnormal object, exotic vacuum object, long-living autonomous object, spherical plasma formation, etc. So many experimental data were collected during two-century researches that there is no necessity to carry out additional experiments to research of AGO properties.

At present, there is an understanding that the circulating light takes place in a wide class of AGO of various sizes from a fraction of millimeters till tens meters, with various lifetime from microseconds till ten seconds, with various stored energy from a fraction of Joule till mega Joules. Besides, the shell of the circulating light can consist of not only compressed air but also various other gases or liquids.

Of course, it is impossible even to list all attempts to obtain Ball Lightning in a laboratory. Therefore, we will focus on the most revealing and informative from our point of view. These are our attempts in 2008 to study luminous silicon balls with anomalous properties and thea lifetime of up to 8 s, attempts to study continuously luminous formations during the discharge of wet electrodes in 2018, the most interesting attempts of Russian scientists described in the collection "Ball lightning in the laboratory" [89] and the attempts described in Barry's book "Ball lightning and Bead Lightning" [3]. It should be borne in mind that our goal is not to describe existing attempts to reproduce Ball Lightning in a laboratory but to analyze a presence of the circulating light in observable autonomous objects.

In our opinion, almost all luminous objects known to us with anomalous properties can be explained by the presence of circulating light in one form or another. Perhaps in some cases, there may be another explanation we do not know. We only pay attention that anomalies can be explained taking into account the action of the circulating light.

Before presenting the material, it was necessary to solve the problem of the sequence in which it is advisable to present. We admire the experiments of researchers in those days when there was no modern scientific equipment, and the voltage was still measured in a number of galvanic cells connected in series. But presentation in chronological order has a drawback because we need to repeat the same comments several times. Therefore, a different order was chosen in which attempts are grouped into sections, in each of which the circulating light has its specificity in production methods and properties.

Besides, the most convincing attempts, in our opinion, should be considered first to show the reader that circulating light does exist not only in natural Ball Lightnings.

By a simple analysis of the experimental data reported by the researchers, we have obtained additional information about the circulating light. Whether to consider this information as new phenomena or as phenomena that can be explained based on existing laws is up to the reader to decide.

In this chapter, we will try to explain the nature of those phenomena that contribute to an appearance of the circulating light in the anomalous objects. It should be noted in advance that our theoretical study is again based on the experimental data. In fact, we are trying to explain the physical phenomena responsible for the observed effects. Therefore, as in the case of Ball Lightning, our explanation should lead to results that are known in advance. Perhaps this circumstance may give additional arguments that the theory of the circulating light is correct.

At the same time, it is difficult to expect that we will give comprehensive explanations about the nature of all anomalous objects. For centuries, scientists could not find an explanation of the most famous object in the form of Ball Lightning. We hope that taking into account such a new phenomenon as circulating light will allow in the future to more fully describe the physical nature of objects that are considered abnormal.

Back in school, they explained to us the difference between science and technical progress. Science gains new knowledge like a miner gets coal. Technological progress uses existing knowledge to create useful things. For example, modern smartphones are based on knowledge from the mid-20th century when transistors were invented. As a result of technological progress, the technology of making transistors has improved so much that the modern smartphone is superior to the supercomputer of the time when attempts were made to fly to the moon. New knowledge can be obtained from the explanation of phenomena that have not yet found their explanation. We cannot explain why theoretical physicists do not interest in the anomalous phenomenon in the form of the anomalous behavior of luminous balls. It is hard to assume that their knowledge insufficient. Not being such, we nevertheless decided to analyze these phenomena.

This does not require laboratories equipped with the latest technology with unique measuring instruments. It is enough to shoot a video on a regular smartphone using a simple installation on the desktop.

To this end, we first repeated the experiments available on the Internet and made sure that the anomalous behavior of the luminous balls does indeed take place. By creating various conditions, we were able to analyze properties of the luminous balls.

Let us begin our analysis with those objects in which a presence of circulating light is obvious and not in doubt.

The Nature of Circulating Light in Bouncing Glowing Balls

Having studied the information about anomalous luminous objects (AGO), in books [1-4], video on the Internet, we concluded that it is advisable to start the study with the most interesting and convincing experiments for us, which are easy to repeat and the abnormal properties of AGO in these experiments are obvious.

A distinctive feature of such AGOs is their lifetime. Unlike other AGO, whose lifetime is fractions of a second and whose anomaly is associated with the traces left by them, the lifetime of the considered AGOs can reach 8 s. During this time, provided that there is a video of the AGO, it is possible to study in detail their anomalous behavior.

Quite unexpectedly, the circulating light manifested itself in the area where a shell of compressed air is absent, and the light circulates along the surface of aglass ball. In this case, the light arises and is maintained due to an exothermic chemical reaction, as a result of which silicon transforms into a transparent glass on the surface of the AGO.

Brazil scientists reported in 2007 that glowing balls (AGO in our interpretation) can appear at conventional arc discharge with small pieces of Si wafer introduced in the discharge gap [90]. Unlike conventional sparks that accompany an arc discharge, AGOs have unusual properties. At first, their lifetime is essentially greater than that of the sparks and can achieve eight seconds. In second, they jump in the process of their moving, can bypass obstacles, and penetrate through splits which width is smaller than their diameter. In the opinion of the scientists, these AGO properties remind a behavior of Ball Lightning. A video film about the behavior of these objects was presented. True, it can be understood from this film that AGOs have an abnormally long lifetime, but it is impossible to make sure that AGOs have the listed properties. Then the same experiment was repeated by American, Spanish, and Holland researchers. However, no attempt has been made to explain the observed anomalies by anyone. Only exclamations of surprise are heard in these videos.

Unlike other experimenters, we know how objects in which the light circulates should behave. Therefore, we had the opportunity to pre-design the special experimental conditions under which we could predict the behavior of the AGO. We can say in advance that our expectations were fully justified and we had no doubt that intense light is circulating in AGO.

Knowing the properties of circulating light, we decided to check whether there is circulating light in such balls. Like Ball Lightning, AGO has to react to obstacles in different ways if the thermal conductivity and heat capacity of the obstacles are changed. Unlike Ball Lightnings, AGOs have significant weight and are subject to gravity. They fall downwards like ordinary bodies. Therefore, it was decided to check how they bounce off different horizontal surfaces at contact with them.

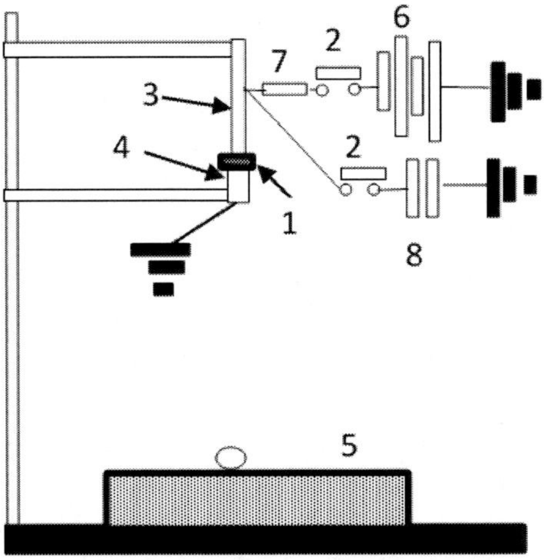

Figure 18. Installation for producing luminous silicon liquid balls. 1 silicon plate of 3 * 3 * 1 mm. 2 is a switch, 3 is the positive electrode, 4 is the negative electrode, 5 is a horizontal substrate, 6 is DC power supply, 7 is a resistor, 8 is a capacitor.

According to our ideas, the balls should bounce in the best way from the surface that heats up faster when the ball approaches it. The causes of this phenomenon are given in Chapter 1 when the reasons why Ball Lightning can bounce are discribed. The bottom line is that the Ball Lightning heats an obstacle due to radiation. In turn, the obstacle heats the layers of air located near it due to the phenomenon of thermal conductivity. The reflective index of the heated air decreases, and the optically induced force arises directed towards the cold air, that is, from the obstacle. Under the influence of this force, AGO bounces off an obstacle.

Concerning the ball that falls on a horizontal surface, which can be considered an obstacle, it should bounce off obstacles that are heated more strongly by the ball. These are either obstacles with low heat capacity and thermal conductivity (a sheet of paper, a wooden board), or very thin obstacles (foil). On the contrary, the ball should not bounce off obstacles that are not heated by the ball (a thick sheet of metal), transparent obstacles that transmit radiation through them (ordinary plate of glass).

The experiments were conducted in 2008. To check the presence of circulating light, we put together the simplest installation on a regular desk. The results of the experiments are presented in articles [33, 35. There is also described the experimental setup and given its photo. The simplest scheme of this setup is shown in Figure 18.

A piece of a standard silicon plate 1 in a form of a square 3 * 3 mm in size and about 1 mm thickness is placed between two electrodes 3 and 4. The piece is taken from the wafer used in the manufacture of integrated circuits. The surface of silicone plate 1 is covered with a film of silicon oxide, which is an insulator. Therefore, it is not possible to directly heat the plate with the current to a temperature at which it melts. A breakdown of the film is required. To this end, first, a high voltage is applied to the plate from the pre-charged capacitor 8 charged to 1.4 kV.

After this, a current from several car batteries 6 connected in series is passed through the plate. The current value is set by the resistance 7 connected in series with the plate. As a result, plate 1 heated and turned into a liquid in the form of luminous white balls. The balls fell on a horizontal surface 5.

Unlike all known experiments, our experiments were carried out, taking into account the fact that the observed luminous balls with anomalous properties contain the circulating light. In this case, the experimental conditions were chosen in such a way as to make sure that the behavior of anomalous objects is determined by the light circulating in them.

We used the following types of obstacles: paper, heat-sensitive fax paper, wooden board, textolite, foil textolite, metal plate, window glass. All obstacles had a flat surface and were located horizontally on a table perpendicular to the direction of falling balls.

The reader can guess for himself which obstacle provides the best ball bounce.

In these experiments, it was most convincingly shown that the behavior of anomalous objects is determined by the light circulating in them. The following anomalous properties could be observed.

A luminous ball about 1.5 mm diameter when falling onto ordinary A4 white sheet paper, located on the horizontal surface of the table, begins to dance on this sheet for several seconds. The dance is that when the ball touches the paper, it bounces off the paper and rises to a height of about 20 cm. After that, it again approaches the paper, and the process repeats. There are more than two dozen such jumps. At first, the height of the bounces is approximately the same, and then gradually decreases. Of course, the ordinary chilled solid silicone ball cannot bounce off the paper even once.

If thermal sensitive paper for fax was used as a sheet of paper, which darkens when heated to a temperature of more than 70°C, dark spots remained on the paper at the places where the ball bounced off the paper. The same paper, when heated to more than 200 C becomes white. Ten the black spots mean that the paper was heated to a temperature of less than 200 C.

To get an idea of the temperature of the fax paper that the ball heats up when approaching, you can see the rings left by the ball when it burns through the paper in in Figure 19. The outermost ring from the center has the lowest temperature and is black. At higher temperatures, the ring turns white. Even at higher temperatures, the paper burns through and turns black. You can see that the trace of the place to which the ball was approaching is visible on paper in the form of a black round spot. But in the center of this circle you can see a small speck of white. This indicates that the center was heated more strongly than the periphery. In addition, you can see tracks from a rolling ball in the form of caterpillars. Such traces are reported by many researchers of objects with anomalous properties.

Figure 20 shows the traces left by a luminous ball on a heat-sensitive paper used in fax machines. All tracks are numbered. As you can see, the ball makes more than 20 bounces from the fax paper, which has no support. An ordinary ball will not bounce off such paper even once.

Two traces of jumps can be seen in Figure 21. The size of the black spots gradually decreases, which indicates that the size of the AO gradually decreases over time due to a chemical reaction that ensures the presence of the circulating light for a relatively long time.

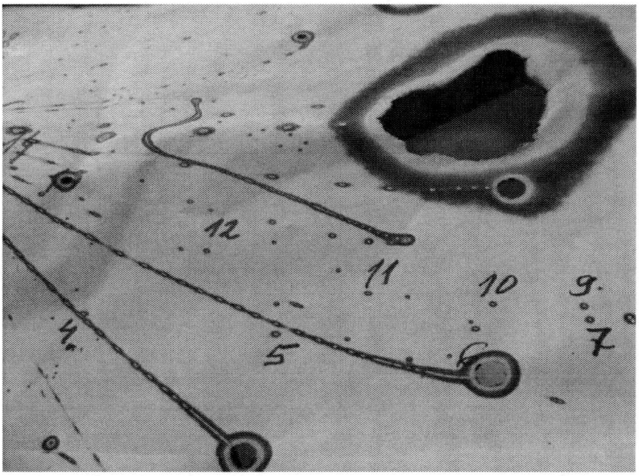

Figure 19. Traces from the collision of the balls with a fax paper. The burnt paper is visible in the upper right corner. The white ring around the hole indicates that these areas of the paper have been exposed to a lower temperature than the burning temperature of the paper. The black ring that surrounds the white indicates that the temperature of the paper was lower when the dark ring formed. Thus, white fax paper, when heated, first turns black and then turns white again.

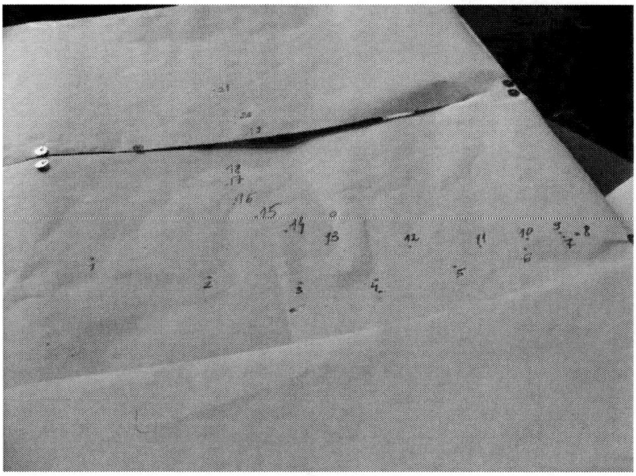

Figure 20. Traces from the collision of the ball with a fax paper without support. Traces left by two luminous ball are numbered. As you can see, the size of the black spots is gradually decreasing. This indicates that the size of the luminous ball is gradually decreasing.

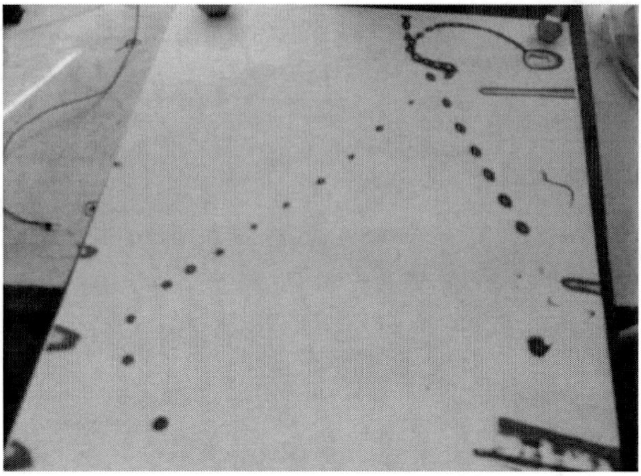

Figure 21. Traces from the collision of the ball with a fax paper.

The same conclusion can be drawn from Figure 22 where the ball bounces off the black cardboard. With each jump, a white spot remains on the cardboard, since at this moment the speed of the ball is minimal, due to a change in the direction of speed to the opposite.

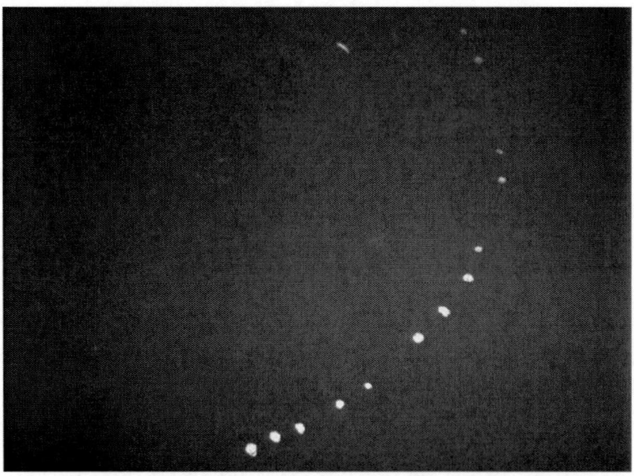

Figure 22. Traces from the collision of the ball with a black cardboard.

Studying the causes of smoke that accompanies the movement of the ball, we concluded that the smoke consists of fine solid particles. Most likely, this is silicon nitride (Si_3N_4), because, as follows from the catalogues, it is a white powder. This reaction heats the drop continuously during its existence. The presence of white powder is confirmed by white traces left by AGO on black cardboard shown in Figure 22.

It can be assumed that the anomalously long lifetime of a luminous ball without a noticeable change in its temperature is due to an exothermic reaction that maintains the high temperature of the ball.

The surface of the ball is silicon dioxide or silica. As is known, silica is an excellent optical medium that is used for the fabrication of optical fibers.

Thus, nature tells us yet another way to form the circulating light, in which the light circulates in a highly heated liquid glass spherical shell. In this case, the light energy is continuously replenished due to the emission of light from excited atoms located in the hot shell itself.

The traces of the ball are shown in Figure 23 when the ball rebounds from copper foil glued to the textolite. The foil has time to heat up to a temperature at which the gradient of the refractive index arises, sufficient to reduce the speed of the approaching ball to zero and give it the speed in the opposite direction.

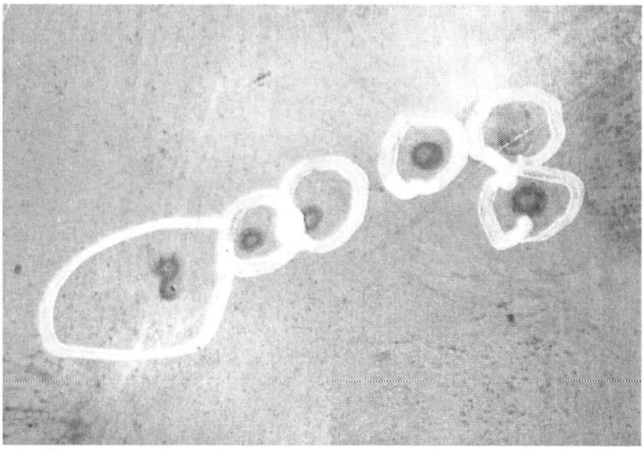

Figure 23. Traces of jumps on foil textolite. The traces are circled with a felt-tip pen.

However, the situation changes when the ball bounces off conventional window glass. The glass is transparent, so it is slightly heated by the radiation of the ball. The light enters through the glass into the substrate in the left Figure 24. A trace of heating of the substrate is visible in the form of a slight darkening. When the ball is in contact with the glass for a long time, its temperature decreases to such an extent that it breaks up into small pieces, which are visible in the right Figure 24.

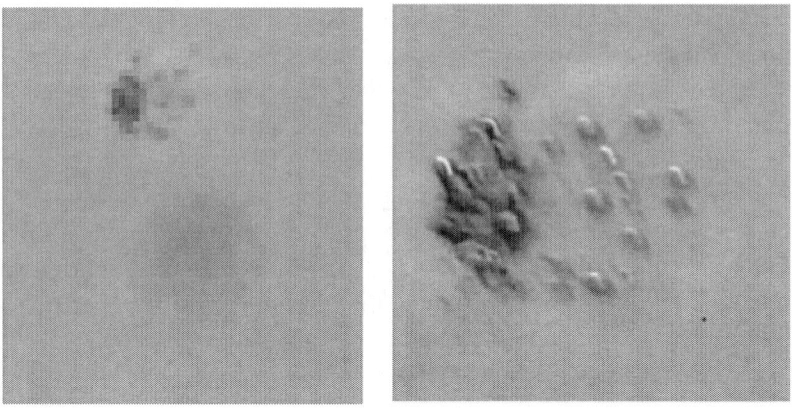

Figure 24. Spatter from a liquid ball that falls on a plate of transparent glass. The darkening of the substrate under the glass from heating by radiation from the falling ball is seen in the left figure. After cooling, the ball crumbled into pieces, which are visible in the right figure.

In addition, we could verify the following.

The ball does not bounce off a thick metal object, but it bounces off a thin metal foil.

The ball bounces off a smooth wooden board, but bounces badly off a rough board.

The ball does not bounce well on rough fabric

The ball does not bounce off coarse sandpaper.

The lifetime of various balls is in the range from 2 to 8 seconds.

A simple conclusion follows from these experiments. The greater the air temperature between the ball and the obstacle, the better the ball bounces off the obstacle.

A more detailed description of all processes connected with luminous balls and additional photographs are presented in articles [33, 35].

Why Circulating Light in Ball Light Violates Conventional Laws of Motion

Watching a video with bouncing balls, you immediately notice that the balls do not jump, like ordinary steel balls on a steel plate or balls from a tennis table. It seems that AGOs don't want to abide by the laws to which they must obey. They rise more slowly and lower more slowly. However, none of the hundreds of thousands of viewers tried to explain this obvious unusual phenomenon.

To be convinced of its unusualness, it is enough to gauge by eye the height of the balls bounce (about 20 cm) and the time between two bounces. To estimate the time, you do not need to use a stopwatch. It is enough to count the number of frames between two jumps and take into account that the time of one frame is 40 ms. Moreover, special accuracy is not required since the discrepancy between the actual time between bounces and the estimated time, which is determined in the initial lessons of physics in a secondary school, is about 2 times.

It is easy to verify that the time of raising the ball and the time of falling of the ball are the same, and the time between two consecutive touches of the table is approximately two times longer than the theoretically estimated time, provided that the balls are liquid. Such a situation can occur if the gravity force acting on the ball is four times less than the usual gravity force, equal to mg, where m is the mass of the ball, g is the gravitational acceleration.

The acceleration of the ball can be reduced if it is assumed that it is hollow and the buoyancy of Archimedes must be taken into account. The acceleration of the ball can be reduced by assuming that it is hollow to such an extent that the force of Archimedes is 3/4 of the force of gravity. However, in this case, the force of air resistance, determined by the Stokes law, takes effect. It was shown that after the first jump, the speed of approaching the ball to the table is so small that the condition that the time of raising the ball and the time of lowering it should be approximately the same is not fulfilled [35].

Under modern concepts, the movement of the AGO is determined by the following three known types of forces. This is the force of gravity, the buoyant force of Archimedes, and the Stokes force arising from air resistance. These forces are well studied, their magnitude is known, but any combination of these external forces cannot explain the motion of the AGO under any assumptions about its density, weight, and internal structure. Such a movement should be recognized as abnormal.

At the same time, such a motion can be explained if we take into account the action of one more fourth type of force, namely, optically induced forces (OIF). As shown, such forces begin to manifest themselves at extremely high intensities of light, which creates a circulating light. The presence of circulating light in Ball Lightning explains the uniform motion of Ball Lightning in the Earth's atmosphere at a constant speed, despite the presence of the Stokes force.

This explanation is based on the fact that the density of the air on the front hemisphere of the ball is greater than on the back one. Besides, the air temperature on the front hemisphere of the ball is less than on the back. These facts contribute arising an optically induced force applied to the ball, which compensates for the Stokes force. A detailed analysis of the influence of the optically induced forces on the movement of the Ball Lightning is given in article [35]

A combination of 4 types of forces enables us to have various types of AGO movement. Note that the OIF value is proportional to the energy $\mathcal{E}$ stored in the AGO. The energy is chosen to ensure the required type of motion. The simplest and easily understood combination of forces is as follows. Gravity force is combined with the Archimedes force so that the total force is 4 times less than gravity force in a vacuum. With this combination, the AGO should be hollow. The Stokes force is combined with the OIF so that their total force is zero. In this case, the movement of AGO resembles the movement of an ordinary body in airless space, in which the acceleration of gravity is 4 times smaller than on the Earth. The time of one full jump for an ordinary steel ball is inversely proportional to the square root of the acceleration of gravity g. Since g is reduced by 4 times, the time of one full jump increases by 2 times.

Thus, circulating light exists in AGO whose properties could not be explained based on the known laws of physics. Such objects were considered anomalous. Accounting for the circulating light allows us to explain the anomalous behavior of such objects. In this case, the laws of physics known in the 19th century are enough.

We can conclude that the circulating light is not an oxymoron, not a game of imagination, but an objective reality that exists in nature. If this reality is taken into account, then it is possible to explain phenomena that had no explanations, or rather all explanations were reduced to a single word "anomalous."

Sonoluminescence and Circulating Light

The essence of the phenomenon of sonoluminescence is that a brightly luminous ball appears in a center of the spherical standing acoustic wave exciting in ordinary cold water. The reasons for the appearance of such a ball are known. At that moment in time, when the pressure produced by the standing acoustic wave is minimal, conditions arise for the discontinuity of the water. As a result, a cavity appears, which begins to fill with water vapor and gases dissolved in water. Under the action of the surface tension forces, the cavity acquires a shape of a ball, whose surface is minimal among that of all other bodies of the same volume.

With a subsequent increase in the pressure of the acoustic wave, as well as under the action of the forces of surface tension of water, a diameter of the ball begins to gradually decrease. In this case, the gases inside the ball are compressed. As a result of the adiabatic compression, the temperature of the gases increases. It is shown that the speed of diameter decreasing can exceed the speed of sound in water by 3–4 times. In this case, a shock layer is formed at an interface between liquid and gases, the temperature of the gases in which can reach 5000 - 6000 degrees Celsius. This process is repeated with the frequency of the acoustic wave, which is usually several tens of kilohertz.

However, this phenomenon contains the following strange moments, for which there is currently no explanation.

1. When inert gases are dissolved in water, the radiation power increases hundreds of times.
2. The observed spectrum of radiation corresponds to the radiation of a black body, heated to the temperature of over 100,000 degrees Celsius.
3. The radiation ends in a flash lasting from 10 to 100 picoseconds and the power of about 10 mW.
4. When the water temperature decreases from 40 to 20 degrees, the radiation power increases several times.

These circumstances indicate that the currently existing explanation of the phenomenon of sonoluminescence is incomplete. At the same time, these circumstances find a natural explanation in the assumption that there is the circulating light inside the cavitation bubble. This is the usual white light radiated by heated gas. Such light, circulating in a thin spherical film of highly compressed gas in all possible directions, is considered in the optical model of Ball Lightning [1-37]. The film of a highly compressed gas, the reflective index of which is greater than that of the surrounding space, is a 2D lightguide with a nonzero curvature. Such a lightguide prevents the circulating light from radiation into the surrounding space. In turn, the circulating intense light compresses gases in the film, in which it is propagating due to the optical electrostriction pressure. Thus, there is a small Ball Lightning inside the cavitation bubble.

Note that the concentration of gases in the place of the appearance of a new cavitation bubble is increased in comparison with the concentration of the same gases in other places. This is due to the fact that the gases left over from the previous bubble did not have time to diffuse in a fraction of one period.

The bright sonoluminescent flash usually has a more or less smooth spectrum, without any separate spectral lines. This spectrum grows steeply towards the violet side and is approximately similar to the spectrum of radiation from an absolutely black body with thetemperature of the order of a hundred of thousands of Kelvin. It was the spectrum that became the main stumbling block when trying to explain the phenomenon. If sonoluminescent light has a thermal origin, then it is necessary to explain how an ultrasound wave can heat water to such temperatures. If high temperatures have nothing to do with it, then what is the general origin of light? The spectra of all inert gases have pronounced lines that are absent in the observed radiation. The question arises where does the radiation come from?

Before answering this question, let us consider a simpler question - the possible causes of radiation. It is known that the bubble radius changes by 2 μm in the last 2 nanoseconds before the disappearance of the pulse. In this time the velocity of the radius decrease is about 1 km/s.

In accordance with these rough estimates, the velocity of motion of the boundary between the liquid and the cavity inside the bubble is 3 times higher than the velocity of sound in gas at normal pressure. Researchers estimate this speed at 3-4 M, where M is the velocity of sound. Thus, a decrease of the radius is accompanied by an appearance of a shock wave. It is known that the gas temperature at the leading edge of the shock wave increases by a factor of M^2, that is, by a factor of 9-16 in the case under consideration. Thus, the temperature of the compressed gas is in the range of 3000 - 5000 Kelvin and such a heated gas radiates light like any heated body. This explanation is given in [38].

As is known, the radiation intensity is proportional to the fourth power of the temperature of a radiated body. Thus, the presence of inert gases makes it possible to increase their temperature during compression, which leads to a significant increase in the radiation intensity. Regardless of the processes responsible for the heating of gases, we must conclude that the observed light is emitted by hot excited atoms and the spectrum of this light contains bands.

Let us now consider the most mysterious phenomenon associated with a significant decrease in the wavelength of the radiated light and a disappearance of bands in the spectrum of the observed radiation, although these bands are certainly present in the spectrum of excited atoms.

Let us show that the circulating light takes part in this phenomenon. The simplest explanation is that the light radiated by atoms circulates in the rim of the cavitation bubble, the radius of which decreases with time. A decrease in the radius is accompanied by a decrease in the wavelength of the circulating light. In this case, the wavelength λ decreases in proportion to the radius R [36].

Ultimately, the radius of the shell of the circulating light will be reduced to such an extent that radiation loss became dominant. The circulating light becomes unstable and disappears. In this case, a flash of radiation intensity is observed as radiation losses increase exponentially in time in this case. A similar picture was considered early at the disappearance of ball light.

As for the disappearance of bands in the emission spectrum of atoms, this phenomenon can be explained in one word - blurring. This means that radiation injected at one frequency into the circulating light is compressed, as a result of which this frequency is increased. However, the circulating light has radiation losses. As a result, parts of radiation are emitted at different times with different frequencies that turned out to be increased relative to the initial frequency by different values. Thus, radiation entered at a fixed initial frequency is emitted into free space in the form of parts with different energies and different frequencies at different times. The spectrum of such radiation will be continuous. The spectrum of a cavitation bubble is a superposition of such spectra.

Based on the presented picture, it is possible to draw some conclusions about the appearance and behavior of the circulating light in the cavitation bubble. For purposes of analysis, let us assume that a duration of the entire process of the generation and conversion of radiation is equal to the duration of the emitted pulse and is about 200 ps. During this time, the radiation frequency has increased by a factor about $k = 2$. A burst of radiation occurs at a wavelength of 0.2 μm. Then, in 200 ps, the bubble radius decreases from 1.2 to 0.6 μm. In this case, the velocity of decreasing the radius is equal to 3 km/s. It is close to the value mentioned in [38].

We can assume the following process of arising and modification of the light in a cavitation bubble. The light is radiated in the edge of the shock wave where the gas temperature is about 5000 C. A part of the radiated light radiates in the external water. Another part is circulating in the spherical shell near the surface of the bubble as is shown in Figure 25.

There is a layer of cold air 2 immediately near the border with the water, which is cooled by the cold water. The reflective index of this layer is smaller than that of the shell. This prevents radiation of the circulating light in surrounding water, reflective index of which is greater than that of the shell. This assumption makes it possible to explain the doubling of the frequency of the circulating light. For this, it is necessary that a part of the circulating light remains in the shells 3 during the decreasing the shell radius by a factor of 2.

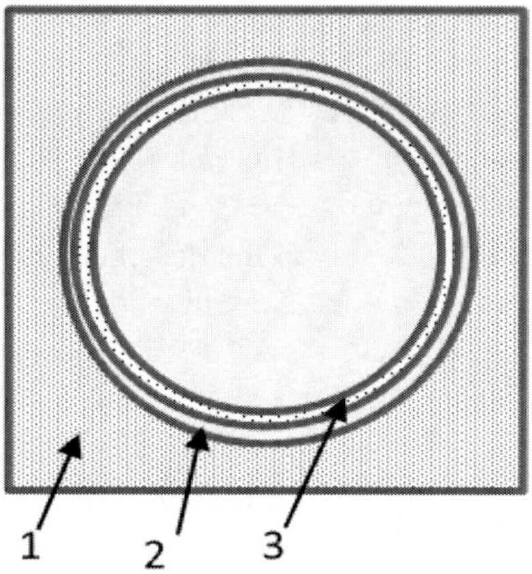

Figure 25. Cavitation bubble in liquid where 1 is the water, 2 is the spherical boundary between the liquid and cavity, 3 is the spherical layer of the hot compressed air where the light is circulated in all possible directions.

There is another way of the same explanation for the increase in the frequency of light radiated by the cavitation bubble. We present it with the aim of a more complete understanding of the processes occurring in the cavitation bubble.

Every photon emitted by an excited atom in any direction, in any case, crosses a side surface of the bubble. This boundary is also the boundary of two optical media with different reflective indices. A part of the light enters through the border; the other part is reflected. A ratio between these parts depends on the angle of incidence on the boundary and the polarization of the light.

Unlike a conventional boundary, our boundary is moving, and the boundary moves towards the incident light. In accordance with Doppler's law, the frequency of the reflected light increases.

In this case, an increase in frequency is proportional to the speed of the boundary movement, that is, the speed of the decrease in the radius. The reflective beam experiences multiple reflections, and therefore its frequency increases many times by $\Delta\omega$, the value of which increases with an increase of the velocity. With each reflection, a fraction of the beam's energy is radiated into free space.

Figure 26 shows spectrums of light that are radiated by atoms in different time instants t_0-2Δt, t_0-Δt, and t_0. As is seen, the earlier an atom emits light with respect to the time t_0 when the spectrum is observed, the more significant is the shift of the spectrum of this atom towards short wavelengths. Figure 26 shows only three time instants when atoms radiate the light. As a result, a number of observed lines increases by three times. Since atoms radiate light continuously, the number of such moments is much greater. As a result, a number of observed spectral lines is so great that it is impossible to distinguish one line from another. Therefore, the observed spectrum is continuous.

Thus, beams with different frequencies, arising after multiple reflections from the moving boundary, shift the individual lines in the spectrum radiated by excited atoms. Over time, new photons appear, bouncing off the boundary, which is moving at a higher speed.

From the picture under consideration, it becomes clear why the spectrum of light emitted by the cavitation bubble is uniform, and the wavelength of this light gradually decreases with time.

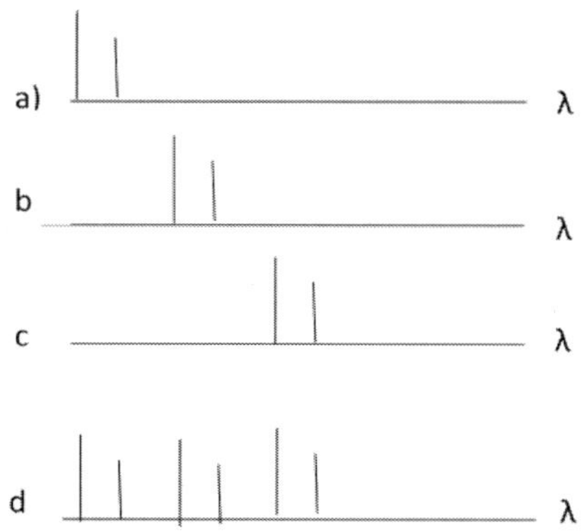

Figure 26. Observed spectrums of radiation from the circulating light at t=t_0.

Spectrum radiated atoms at t=t_0-2Δt and shifted in short wavelength due to the compression of the light in time 2Δt.

Spectrum radiated atoms at t=t_0-Δt and shifted in short wavelength due to the compression of the light in time Δt.

Spectrum radiated atoms at t=t_0.

Total spectrum observed at t=t_0 as a sum of spectrums radiated at t=t_0-2Δt, t=t_0-Δt, t=t_0.

Thus, there are two explanations for an increase in the frequency of the circulating light with a decrease in the radius of the circulation. One of them is considered above and is associated with the fact that number N of wavelengths in the circulating light is preserved with decreasing of the radius of the ball.

The second explanation is based on the Doppler effect, according to which the circulation of light is based on multiple reflections. As the radius of circulation decreases, each reflection is accompanied by an increase in the frequency of the reflected light. Both explanations give the same result. However, the first explanation applies to a conservative system with a constant total energy. A cavitation bubble is not such a system. Therefore, the first explanation can only serve as a hint of what will happen as the bubble radius decreases.

The wavelengths of the circulating light change in a direct proportion with the radius. The change can be significantly when the radius of the bubble is small. The spectrum of the observed radiation from the cavitation bubble, located in the short-wavelength region, does not correspond to the spectrum of the radiation of an object heated to a temperature of hundreds of thousands of degrees. The observed radiation in the short-wavelength region of the spectrum corresponds to ordinary radiation, the wavelengths of which are decreased, when this radiation is compressed at decreasing radius of the cavitation bubble.

With decreasing radius, the speed of movement of the boundary between water and gas increases. In this case, the frequency shift $\Delta\omega$ increases when light is reflected from the moving boundary. In addition, the time interval between adjacent reflections is decreased.

The burst of intensity near the trailing edge of the pulse can be explained as follows. When the radius of the ball becomes comparable with the wavelength, the approach based on the geometric optics becomes incorrect. The wave approach is required. It is known that the losses in the ball where $R/\lambda<2.5$ increases abruptly and all light energy that was stored in the ball of greater radius radiates abruptly.

It is known that the sonoluminescence does not appear in hot water. Moreover, if the water temperature was lowered from 40° C to 0° C, the brightness of the luminescence increased a hundred times.

In addition, we will keep in mind the following known facts. Gases in cold water dissolve better (it is known that there are more fish in cold seas, since there is more oxygen in cold water, which is breathed not only by people, but also by fish). In addition, gases whose molecules contain one atom (inert gases) dissolve in water better than gases whose molecules consist of two atoms (oxygen, nitrogen). The water is an excellent solvent.

An even closer analysis shows that the surface tension forces of water depend on temperature. The greater water temperature the smaller the index of surface tension. In addition, the adiabatic indices for all inert gases are approximately the same and differ markedly from that of the air. All these seemingly unrelated oddities are naturally explained if we take into account that the circulating light is directly involved in this anomalous phenomenon and the following process take place.

With regard to the beneficial effect of cold water on radiation intensity, the surface tension coefficient increases as the water cools. The surface tension coefficient η=69.56 10^{-3} at T = 40 C and increases about ten times to η=75.64 10^{-3} at T = 0 C. The shock wave velocity increases with an increase in the surface tension forces and, therefore, with an increase in the surface tension coefficient. In this case, the index M increases. This, in turn, leads to an increase in temperature at the shock front and an increase in the radiation intensity.

When the integrity of water breaks and a cavity is formed, this cavity, under an action of the surface tension forces, turns into an empty bubble. During the growth phase of the bubble diameter, water molecules and molecules of dissolved gases are drawn into this spherical cavity. Moreover, the colder the water, the less water molecules evaporate into the cavity and the more dissolved gas molecules are drawn into the cavity. Thus, we have explained that the inert gas molecules in the cavity play an essential role in the formation of light in the cavity.

We have explained the entire set of seemingly unrelated conditions favorable for the occurrence of the phenomenon of sonoluminescence. This phenomenon is most pronounced in water, since water is a good solvent. The beneficial effect of lowering the water temperature is explained by the fact that the surface tension force and the solubility of gases in water increase. The favorable effect of inert gases is explained by the fact that the adiabatic index of such gases exceeds the adiabatic index of oxygen and other gases. An increase in the adiabatic exponent contributes to a significant increase in the gas temperature during adiabatic gas compression.

The shift of radiation to the region of shorter waves is associated with a continuous increase in the frequency of the circulating light when it is compressed with a decrease in the bubble size. The absence of characteristic bands in the external radiation, which are present in the emission spectrum of inert gases, is associated with the smearing of external radiation, since it is the sum of radiations at various stages of compression. As in the case of Ball Lightning, an explanation of many different unrelated factors suggests that such an explanation corresponds to reality.

The extremely high temperature of the body emitting short-wavelength light, determined on the basis of spectral analysis and Wien's law, led to attempts to use this phenomenon to carry out thermonuclear fusion reactions by means of sonoluminescence [43, 44]. It is known that for such a reaction, the plasma temperature should be several hundred millions degrees. From our consideration it follows that such attempts are in vain because short-wavelength radiation is a result of the compression of the light radiation rather than the extremely great temperature.

Circulating Light at Continuous Gas Discharges

Very pictorial and indicative experiments were carried out by the famous French physicist Gaston Plante, who invented the lead-acid battery in 1859. This battery was the first rechargeable electric battery released for commercial use. This type of battery is widely used in all modern cars. He is a unique inventor whose invention has been used for nearly two centuries in a number of many billions of copies.

Besides, he invented a device that, in modern terminology, can be called a step-up DC transformer. This was at a time when the voltage was still not measured in Volts but was measured in the number of 2-Volt galvanic cells connected in series. Using his transformer, he managed to get a direct current of 4kV voltage, which was unthinkable at that time. Having such a high voltage source, Plante was able to study electrical discharges between parallel metal plates of a capacitor [3]. We believe that not many modern inventors can invent such a device.

When the voltage was applied to the plates, that were covered by a filter paper moistened with distilled water, he observed the luminous ball occurred between moist areas and would not occur between dry areas. Unlike the voltaic arc, the luminous ball could be observed without a mask. The ball moved about the surfaces randomly but remained between damp areas. The ball existed until the voltage source could no longer support the discharge. Plante speculated that the discharge is connected with the evaporation of the water. He believed that this ball is a small Ball Lightning. However, to date, there is no generally accepted opinion whether this ball is associated with Ball Lightning.

Based on our approach, we can note that there is no voltaic arc since the glow of the voltaic arc is bright and it is impossible to observe without a mask. Besides, the voltage of the voltaic arc at the discharge gap of a few millimeters long is approximately equal to the voltage between welding electrodes. The welding voltage is about 20 V rather than several kilovolts that were in the Plante experiments. Plante's experiments have been repeated many times now. However, neither new results nor new explanations were presented.

Luminous balls, the existence of which is ensured by a constant supply of energy from an external source, were observed by many researchers in the early 20th century when the technical capabilities appeared to have high voltage power suppliers. A detailed description of these experiments is given in the book [3]. We will only mention two experiments from this book.

Gezehus studied a character of the 10 kV discharges of alternating current at electrodes of the various forms from various materials. Metal plates, a surface of the water, water columns, and damp sponges were used as electrodes. Between a copper plate and the surface of water separated by an air gap of 2-4 cm thickness, there was the very mobile discharge of constantly changing form. The fiery sphere moved back and forth on the plate with crackling. It was not felt any heat at several inches from the sphere. In the glass bell placed above the device, oxides of nitrogen are gathered.

The nature of physical processes in these experiments is the same as that in the Plante ones. The low-temperature plasma in the discharge gap and intensive light radiation is generated at a passage of an alternating current through the discharge gap. Because of the self–organization of intense light radiation, a ball light arises. Like Plante experiments, the light intensity in the ball light shell was insufficient for the production of a lightguide with small radiating losses. As a result, the ball light disappears after ceasing the discharge. Like Ball Lightning, the ball light radiates the light due to the light scattering of the intense light circulating in the ball light shell. Because of this, no heat is felted near the ball light.

In another experiment, luminous spheres appeared at a negative electrode in the case if two thin metal wires connected with an electrostatic machine were placed on a light-sensitive emulsion of a photographic plate. After separation of such sphere from a brightly luminous negative electrode, the end of the wire became dark, but a small ball moved towards a positive electrode and slowly traversed the plate with extremely muddle way, sometimes stopping at the moment. To pass the distance of 5-10 cm between the electrodes from 1 to 4 minutes is required for the ball. When the ball achieved the positive electrode, the glow disappeared, and a power supply behaved in such a way as if its poles were short-circuited. After developing the plate, the ball trajectory was seen at it. Electrical conductance was only in those places where the ball passed.

The appearance and behavior of the ball can be explained as follows. In that area where the ball is located, an erosive gas discharge takes place connected with decay and evaporation of the emulsion (for example, gelatin on which basis the emulsion is made). The reflective index is maximal in the area. When the matter from which the emulsion is prepared has vanished, the intensity of evaporation falls, and the reflective index in this area decreases. Besides, the ohm resistance of this zone decreases because the emulsion becomes conductive. In this case, the voltage of the power supply is applied completely to the region between the positive electrode and the area. Flowing current heats up the adjacent area located in the direction of the positive electrode. From this area, evaporation begins, and the reflective index in the area increases. Thus, the ball light shifts to this area.

Ought to underline that the ball light exists in time when the plasma exists. The plasma takes place when the gas discharge takes place. This situation corresponds to the Plante, Gezehus experiments, and our experiments with jets of water.

Bearing in mind the properties of the circulating light, we decided to repeat the Plante experiments. Instead of 2000 batteries, we used a high-voltage source from a conventional microwave oven [55].

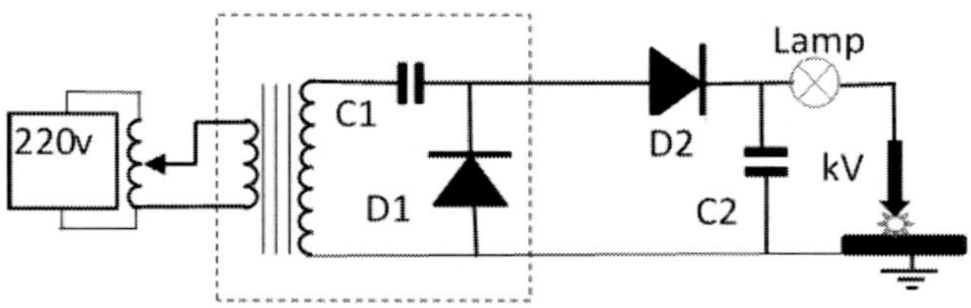

Figure 27. Electric circuit for excitation of AGOs using a continuous gas discharge based on a microwave oven. The dashed rectangle marks the elements inside the microwave oven.

We hope that our experiments will be repeated so that we will give a brief description of our installation. Microwave radiation is created by an electro vacuum device called magnetron, which is powered by a high voltage power supply. The anode of the magnetron has zero potential. Negative high voltage is applied to the cathode through a voltage multiplier consisting of a capacitor and a diode. The power of the microwave oven is changed using pulse modulation produced by a control block. Since we need to have continuous voltage, this block is redundant. A modified scheme for our purpose is shown in Figure 27.

The components bounded by the dotted rectangle are in the microwave. To the left of the rectangle is an autotransformer connected to a regular 220-volt network. The autotransformer allows us to change the voltage supplied to the primary winding of a step-up transformer in a microwave oven. Capacitor C1 and diode D1 are located in the microwave oven. They rectify the high-voltage alternating voltage and multiply DC voltage by 2 times. Since the high voltage is supplied to the cathode of the magnetron, and the magnetron anode is grounded, DC voltage in the microwave oven is negative.

In our experiments, the voltage at the electrode was chosen positive relative to the ground. Therefore, the diode D1 is rotated 180 degrees.

Three components shown to the right of the dotted rectangular are located outside the microwave oven. Diode D2 and capacitor C2 enable to exclude pulsation of DC voltage. The parameters of the capacitor C2 are the following. C=100 mkF, Umax=5 kV. One or several incandescent lamps limits maximal discharge current. As is known, the resistance of the incandescent lamp is nonlinear. The resistance of the cold lamp is smaller by 10 times than the resistance of the same lamp at conventional brightness. The lamps increase the inner resistance of the voltage source.

By observing the glow of the lamps and their flickering, one can roughly judge the discharge current and its pulsations.

Before turning on the installation, the autotransformer slider must be in the zero position. Otherwise, the diode D2, through which the capacitor C2 is charged, may be damaged by the large initial charging current.

With this simple installation, we were able to obtain continuous gas discharges in various situations, which are presented below in the photos with a brief explanation of the conditions under which these discharges were obtained.

Firstly, we repeated the experiments of Plante. The right-hand photo in Figure 28 shows a glowing ball between two wet fabrics that wrap metal discs. Fabrics were moistened in a saturated solution of ordinary table salt to increase their conductivity. In the left-hand photo, two luminous balls are visible adjacent to the upper and lower fabrics. The purple patch between the balls is a gas discharge in a plasma consisting of ionized water atoms and sodium chloride. However, near wet fabrics, the gas temperature decreases due to the evaporation of water from the fabrics. As a result, water vapor condenses into a liquid in the form of a ball or spherical film. Recombination of plasma is accompanied by radiation of white light into the shell of the ball.

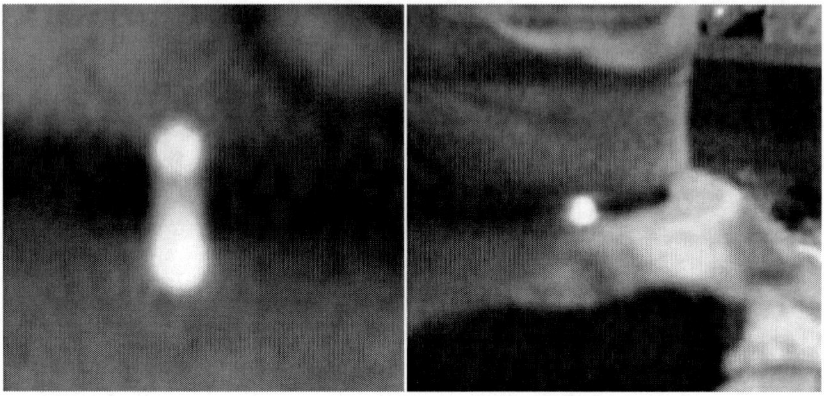

Figure 28. Luminous white balls between wet fabrics at the discharge between them. Two white balls and a week luminous gap between balls are seen in the upper photo. We believe that it is plasma at gas discharge.

The reflective index of water is 1.35. The shell of the luminous balls consists of a water film that is a transparent optical medium, in which white light arising from plasma recombination, can circulate and can be accumulated. A discharge current between the upper and lower electrodes is about 100 mA. The current flows through the balls or the plasma inside the balls. The specific resistance of water is 50 Ohm mm. Therefore, the main voltage drop occurs in the middle gap where plasma is formed. As in Plante's experiments, luminous balls move into wet areas. At the same time, dry areas are not carbonized, which indicates that their temperature is less than 300 degrees.

In the right-hand photo in Figure 28, the diameter of the luminous balls is equal to the distance between the tissues. The question arises - where does the white light come from? A temperature of several thousand degrees is required for a hot body to radiate white light. It remains for us to assume that a gas discharge occurs in the center of the spherical shell, and the temperature at the surface of the fabrics decreases due to the evaporation of water.

Besides, Plante observed luminous objects during a discharge between a metal electrode in the form of a rod and the surface of the water. We also repeated these experiments. Positive voltage was applied to the electrode. The objects arose when the distance between the tip of the electrode and the surface of the water decreased.

A glowing ball during a discharge between a tungsten electrode and the water surface is shown in the Figure 29. Since the thermal conductivity of tungsten is small, the electrode glows dark red. The evaporation of water provides a relatively low temperature of the ball. However, the ball glows white due to the light circulating in its shell, which accumulates light arising due to the recombination of the plasma resulting from a gas discharge. A similar situation was considered early at the accumulation of light in the shell of liquid silicon due to an exothermic reaction.

Figure 29. Luminous balls between the surface of the water and the vertical tungsten electrode. The temperature of the end of the vertical electrode is great because there is no evaporation that is accompanied by the absorption of heat.

In Figure 30, the water surface is replaced by a damp cloth in contact with a metal surface. In contrast to the previous experiment, the upper electrode heats up, since it cannot be cooled due to the evaporation of water. The presence of circulating light in the ball is evidenced by the fact that the brightness of the ball during the discharge process is not constant in time. Sometimes the ball flashes with bright light, which indicates that at this time the circulation conditions were violated and the light burst out. A photo of the frame from the video, where the brightness of the ball has increased significantly, is shown on the right photo.

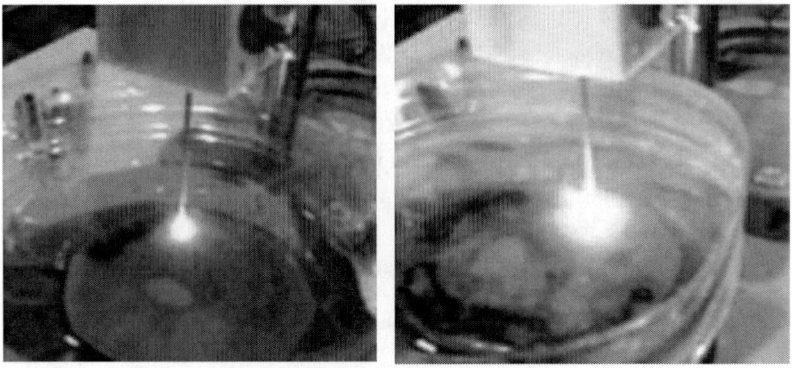

Figure 30. A luminous ball between the surface of the wet metal and the vertical tungsten electrode is shown on the left photo. The flash of the same ball selected from a video is shown in the right photo.

The same character of the luminous balls can be observed at discharge between a thin jet of water from the syringe needle and the surface of the water as is shown in Figure 31.

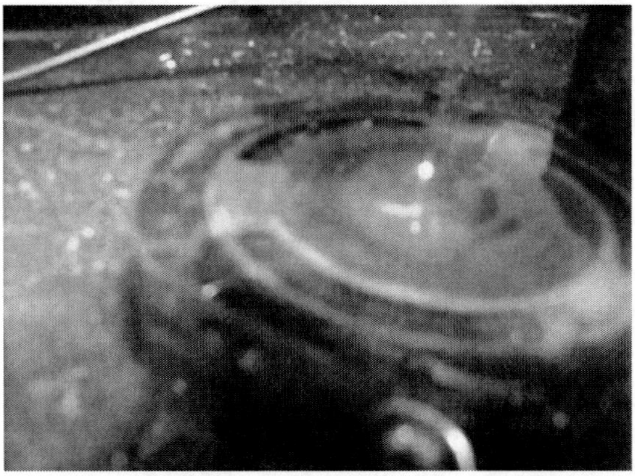

Figure 31. Glowing ball between the surface of wet metal and a vertical tungsten electrode. The weak glow of the ball allows you to make sure that it has the correct spherical shape.

Typical luminous balls between the vertical jet from the syringe needle (anode) and metal (cathode) are shown in Figure 32.

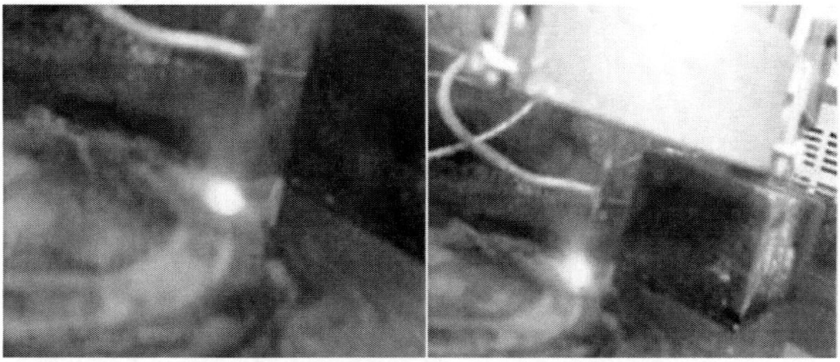

Figure 32. Typical luminous ball between the vertical jet from the syringe needle (anode) and metal (cathode) on the left photo. On the right photo, one of these flashes on the frame separated from the video is presented. We assume that the flash is connected with a disappearance of the circulating light and radiation of the accumulated light in free space.

Figure 33. Luminous ball between the vertical jet from the syringe needle and metal.

The red ball in Figure 33 glows near the exit of the jet from the syringe, where a droplet of water has formed. The light from the lower ball propagates along the fiber in the form of a cylindrical jet of water and is scattered in the droplet in all directions.

It is interesting what the discharge will be between two jets of water if these jets do not intersect. The discharge can occur only in the place where the distance between the jets is minimal. The discharge cannot move along the jets, since with such a movement the air gap increases and the discharge should stop. A luminous ball that appears between non-intersecting jets is shown in Figure 34.

Figure 34. Discharge in a form of luminous ball between two jets of water.

From the photographs presented, it can be concluded that, indeed, during a high-voltage discharge from a source with high resistance, a luminous ball with a clearly defined luminous surface is formed in the coldest place. At the same time, the intensity of the glow is much weaker than that of a voltaic arc, where the glow is explained by the radiation of a substance heated to a high temperature. The weak glow of the balls is explained by the molecular light scattering of the circulating light. The circulating light appears as a result of the gradual accumulation of light, as is the case in the previously considered luminous balls in a form of silicon drops. The glow time of such drops did not exceed 8 seconds, since the substance for these reactions is running out. The duration of the glow of the balls during a gas discharge is limited only by the duration of this discharge, which can exist for an arbitrarily long time if the high-voltage source is powered from the mains.

Thus, the Plante assumption that the luminous balls observed by him belong to the same phenomena as Ball Lightning turns out to be correct. Indeed, these phenomena are united by the light circulating in them.

Circulating Light in a Mixture of Gases

Rather indicative experiments for confirmation of our approach have been carried out by Nauer. His experimental setup is shown in Figure 35 [3]. In a small glass tube 1 with an internal diameter of 5-7 cm, the metal electrode 4 was located. The size of hole 2 at the end of a tube could vary. The second electrode 3 was located near the glass tube. Both electrodes settled down in the glass chamber that could be pumped out. Nauer observed that in a darkened room just after a spark occurred between the two electrodes, a luminous object appeared to rise from the upper hole of the small glass tube. The luminous cloud could last for several seconds.

Nauer investigated the illuminated cloud phenomenon and found different results as the air pressure in the vessel was varied. At low pressures near 0.02 atmospheres large bright clouds formed; at higher pressures near 1 atmosphere smaller and darker clouds formed. Different gases were used: air, oxygen, propane, an air-chlorine mixture, an air-benzene mixture, and argon. Colors, intensity, and cloud dimensions varied.

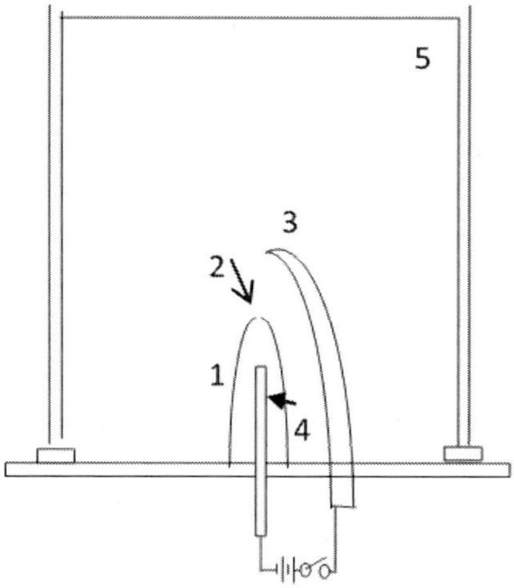

Figure 35. Scheme of the experimental setup used by Nauer where 1 is the glass tube with a small hole 2 on its face; 3 and 4 are electrodes, 5 is the vacuum chamber.

It is important to note that the luminous cloud could be formed in dry air at normal atmospheric pressure. For the cloud to form well, hole 2 at the end of the small glass tube 1, which enclosed electrode 4 had to be quite small. The luminous cloud produced at the normal air pressure was also visible in an undarkened room, although specific details of its form could only be seen in the dark.

It was noted that the luminous cloud rose upwards, apparently as a result of the heated gas from the spark discharge. A temperature increase of 15° C above the room temperature was measured by a thermometer in contact with the luminous cloud. The constriction at the end of the glass tube enclosing electrode 4 improved the formation of the luminous cloud. A small hole was required at atmospheric pressure to produce the phenomenon. Nauer dismissed the possibility that the luminous clouds were oxidation or burning phenomena since such clouds could be produced in inert gases such as argon.

We can explain simply enough the beneficial influence of a small hole 2 in the glass tube 1 on conditions of occurrence of a cloud. For successful ball light formation, it is necessary to increase the time during which the ball light is in a charge gap. All this time the ball light accumulates the light energy in its shell. One of the most often used ways of performance of this condition is the use of an erosive discharge that is accompanied by an evaporation of any substance from electrodes. The evaporation increases the reflective index in the discharge gap up to such a degree that its maximum is located in the discharge gap despite an increase of the temperature in the discharge gap.

Since ball light tends to move to the area with the maximal reflective index, the ball light remains in the discharge gap, and an accumulation of light in its shell takes place. This way has been found out experimentally long ago. However, the reasonable explanation of favorable action of an erosive gas discharge on the formation of autonomous objects did not exist till now.

One more way that allows the ball light to be in a discharge gap for a long time was experimentally discovered by Nauer. Actually, in Nauer installation, a discharge gap occurs in the closed chamber. Only a small amount of gas escapes from the discharge gap, penetrating through a small hole. In this case, despite an increase in the temperature in the discharge gap, the gas density in it remains almost constant, since the gas pressure also increases. Thus, the amount of gas in the tube and, therefore, its density does not change. Since the gas reflective index is proportional to the gas density, the gradient of the reflective index directed from the discharge gap is minimal.

After termination of the discharge, the erosion of electrodes ceases. As a result, the gradient of the reflective index on the axis of small hole 2 in glass tube 1 is directed outside as a part of gas has left the tube through the hole. The ball light penetrates through the hole in the same way as Ball Lightning penetrates through a narrow crack. Nauer has experimentally shown that narrowing of the hole improves conditions of formation of a shone cloud. The same conclusion follows from the presented explanation based on the concept of the ball light.

Nauer experiments connected with burning are rather indicative for confirmation of the concept of an existence of the circulating light. Casually it was revealed, that the small traces of the benzene that has remained in the glass chamber after its clearing occurs a significant influence on an appearance of shone clouds. Experiments were carried out in a closed chamber of 6-8 cm diameter and 50-200 cm long. Between two electrodes placed within the chamber, an electric spark was lit. Shone clouds appeared after the termination of the spark discharge and existed within several seconds. These clouds did not fill the whole volume of the chamber and had a diameter of 2-3 cm. Formation of shone clouds became more probable when the benzene was thoroughly mixed in the enclosed volume of air.

We would like to analyze one very interesting Nauer experiment that enables us to discover the perfect new physical effect connected with a new type of optical nonlinearity. The following unexplained fact has been found out at studying an appearance of luminous clouds at the hydrocarbon concentrations smaller than that of inflammation. The spark discharge gave no visible results in the air-propane mixture if the concentration of the propane was smaller than a threshold of inflammation (about 2.8%). But at concentrations of the propane in a range from 1.4% to 1.8% a yellow-green fireball appeared of several centimeters in diameter. The ball was bright enough and promptly moved in the vessel in about two seconds. After then it disappeared without noise.

All attempts to explain this paradox were unsuccessful. Taking into account the circulating light, the paradox is explained very simply under the assumption that the reflective index in the shell of the circulating light may be increased not only by the air compression but also by drowning in the shell components of the mixture with maximal reflective indexes. The reflective index of propane differs from 1 by Δn, where $\Delta n = 0.00054$. On the other hand, $\Delta n = 0.00027$ for the reflective index of conventional air. It means that a double increase in Δn in the shell of a ball light can be reached by drowning molecules of propane in the shell. For this purpose, significantly smaller the energy of light is required, than that for the compression of air by 2 times. Thus, a small impurity of propane is favorable for the formation of ball lights. The volume occupied by the shell is smaller by several orders of magnitude than that of the surrounding gas. Because of this, a small concentration of the propane is sufficient for the formation of the shell from propane.

Note that at the great concentration of propane, the considered effect disappeared. Our explanation is as follows. The process of absorption of propane in the shell is inertial. Its time constant depends on the concentration of propane and decreases with decreasing concentration, since the mean free path of the molecules increases. Consequently, with a decrease in the concentration, molecules can be drawn from a larger volume. With increasing the concentration, molecules can be drawn into the shell only from nearby layers, which leads to a depletion of these layers. The concentration equalization time, in this case, exceeds the discharge time.

Nauer derived a conclusion that the percentage of combustible gas required to form the luminous clouds is quite small and would appear to be inversely proportional to the molecular weight of the hydrocarbon. In this relation, it is very interesting to check whether the reflective index of gas is proportional to its molecular weight. In the following table, these values are given for gases taken from the handbook [91].

Table 1. Correlation between Δn of gas and its molecular weight

Gas	**$\Delta n\ 10^7$**	**Molecular weight**
N_2	2793	28
NH_3	3750	17
air	2630	40
C_2H_5	6060	29
Br_2	11250	160
HBr	5700	81
H_2	1390	2
H_2O	2254	18
He	2753	8
N_2O	5150	44
HJ	9060	128
O_2	2531	32
Kr	2752	84
Xe	7020	131
CH_4	4410	16
Ne	2716	20
O_3	5110	48
NO	2970	30
CO	3340	28
SO_2	6600	64
SO_3	7370	80
H_2S	6190	34
CO_2	4197	44
F_2	1950	38
Cl_2	7680	70
HCl	4440	36

As is seen from the table, the greater the molecular weight of the gas, the greater, as a rule, its reflective index. In this case, Nauer's conclusion may be reformulated as follows. The smaller a percentage of combustible gas required to form the luminous clouds, the greater its reflective index. Indeed, under our approach, the greater the reflective index in the shell, the better possibility of the shell to confine the light circulating within it. Thus, the great reflective index of admixture is favorable for the production of ball lights.

As can be seen from the table, the reflective index of hydrocarbons significantly exceeds the air reflective index. This explains the favorable effect of an erosive gas discharge, in which hydrocarbon vapors arise during the discharge. Researchers came to this conclusion based on their experiments. We explained this phenomenon by the presence of the circulating light in AGOs.

The effect that luminous clouds become brighter at the addition of benzene or propane is explained by the fact that Δn for these gases is significantly greater than that for other gases used in tests. In this case, the ball light shell can accumulate greater the light that entails the greater light scattering.

In the case where the shell is formed by gases with great Δn, a disappearance of the ball light occurs silently because there is no expansion of gas, and a diffusion of the concentrated impurity occurs only. If the shell is formed by the compressed gas, a disappearance of the ball light is accompanied by a clap because in this case, a sharp expansion of the gas compressed in the shell takes place.

Nauer observed that the luminous clouds were not affected by external electric and magnetic fields. A color of the luminous clouds was variable but directly dependent upon the admixture gas. All colors were possible. The luminous cloud did not appear as a burning phenomenon, but rather as an electrical glow discharge – a soft glow *apparently emanating from the SURFACE of the glowing body*. Mixtures of color, uniform colors, streaks, and stripes were also observed. The light from hydrogen was pale blue, and that from benzene or propane was visible in a brightly lighted room.

The luminous clouds exhibited motion, generally rising, although both up and down motions were also observed. The speed of motion was measured as 0.33 – 10 m/s. No noise was detected from the phenomena at low mixture levels. A slight rustling noise was detected at high mixture values. The cloud dissipated predominantly without noise, but occasionally with a loud noise when the cloud contacted the lid of the container.

One specific trait stands out as being of some significance. The vessel was separated into compartments by a flat disk with a 7 mm central hole placed within the enclosure. The luminous cloud was normally extinguished when it contacted the disk. However, sometimes the luminous cloud appeared to pass through the hole and regain its previous size and shape.

Nauer concluded that there is a definite correlation between natural Ball Lightning and the laboratory-produced luminous clouds formed by a low-density hydrocarbon excitation mechanism. *It was proposed that Ball Lightning has no electrical nature at all.*

Describing these Nauer experiments, Barry concluded that they were most significant since Plante experiments, some 70 years earlier. On our side, we can add that a description of properties of Nauer luminous clouds is at the same time description of the properties of our circulating light. Let us list properties of the clouds that can be easily explained if take into account that they are the circulating light.

Nauer observed that the luminous clouds were not affected by external electric and magnetic fields. In fact, the circulating light consists of light and compressed gas. Neither of them is affected by external electric and magnetic fields.

The luminous clouds radiate a soft glow apparently emanating from the *SURFACE* of the glowing body. Indeed, any circulating light radiates light from its surface because the light is concentrated in the shell in a form of a thin spherical layer.

It is not wonderful that the cloud can penetrate through a small hole and regain its previous size and shape. The process of penetration of the circulating light through a small hole was considered early in chapter one.

The fact that the clouds did not occupy the whole space of the camera is explained by the property of the circulating light to preserve its area of surface constant.

Nauer's conclusion that *a Ball Lightning has no electrical nature at all* is valid completely because the ball lights *have no electrical nature at all* also. Unfortunately, Nauer did not know that the circulating light exists in nature. If he wrote only two words "circulating light," a physical nature of Ball Lightning could be discovered a half-century ago, and, possibly, at present, an existence of the circulating light would have received a general recognition.

Circulating Light at Erosive Gas Discharges

There is a large group of experiments related to attempts to obtain AGO using powerful gas discharges. For example, according to Abrahamson and Dennis [92], Ball Lightning exists due to the oxidation of silicon nanoparticles in the atmosphere. Such particles are formed when an ordinary linear lightning strikes the ground and produces the soil of silicon and carbon oxides. While these concepts cannot explain why Ball Lightning travels against the wind, penetrates through window panes, and catches up a flying airliner, they agree with attempts to create an artificial Ball Lightning. Indeed, as follows from these concepts, an erosive gas discharge is required, in which additional products evaporated from the electrodes must be delivered to the Ball Lightning region.

The increase in the reflective index in the shell of the circulating light can occur not only due to the air compression but also due to drawing the gas components with the maximal reflective indexes in the ball light shell. The light energy for increasing the reflective index at some value Δn is required smaller in the latter case than in the first one. Thus, studying abnormal AGO spectra has allowed us to disclose a new type of the optical nonlinearity in gases.

From the presented consideration follows, that a natural Ball Lightning also can have a shell consisting of various vapors. Really, there is numerous evidence that such Ball Lightnings arise often enough. Occurrences of Ball Lightning are connected often by eyewitnesses with distinctly perceptible odors. More often, these odors were defined as a smell of sulfur and ozone. In several cases, the smell was compared with a smell of the dioxide of nitrogen. All these gases have the reflective index, which Δn is approximately twice greater than that of usual air.

Some observations indicate that a bolt of initial linear lightning can generate Ball Lightning, having struck in any firm substance and having evaporated it. An occurrence of a fiery sphere that slides from a tree prone to a lightning strike is typical enough [2]. The fact of detection of the pitch in residues from Ball Lightning appeared at a strike of linear lightning in a tree is in good agreement with the presented picture

It is also reported about Ball Lightning on the Earth, which exploded after 30 seconds. The ball appeared in an extremely strong thunderstorm. There were tar residues after the explosion. They have the sulfur smell and were so hot that they burned eyewitness hand through 10 minutes after the explosion. From hitting at this substance, his finger became yellow.

There is numerous evidence that Ball Lightning can contain other vapors except for air. For example, Stakhanov describes a case where the hot black mass was observed after the disappearance of Ball Lightning that arises from a linear lightning that strikes into a tree.

Circulating Light at Gas Discharges with Erosion Hydrocarbons

An erosive gas discharge is used widely in experiments presented in [93]. For example, a scheme of the Emelin *et al.* experimental setup is shown in Figure 36. Here a discharge occurs in the discharge gap between electrodes 1 and 2. The discharge gap is formed in a plexiglas plate where hole of 1-2 mm in diameter and 3-4 mm in length is made. Capacitor of 3.2 μF charged at voltage 300-340 V was used as the power supply. The maximal pulse current was 100-150 A. The width of the current pulse was about 1.5 ms.

AGOs or in the terminology of authors long-living shone formations appeared at a gas discharge. Their lifetime was about 1 s, and their diameter was about 1 mm. AGOs left the discharge gap approximately 45 ms after the termination of the discharge. Practically all observable AGOs during their existence made numerous jumps on a surface of the experimental installation covered by a black paper. After decoding the video recording, it was seen that AGOs have almost absolutely elastic impacts. AGOs, as a rule, disappeared suddenly at the next contact with a surface or in a flight.

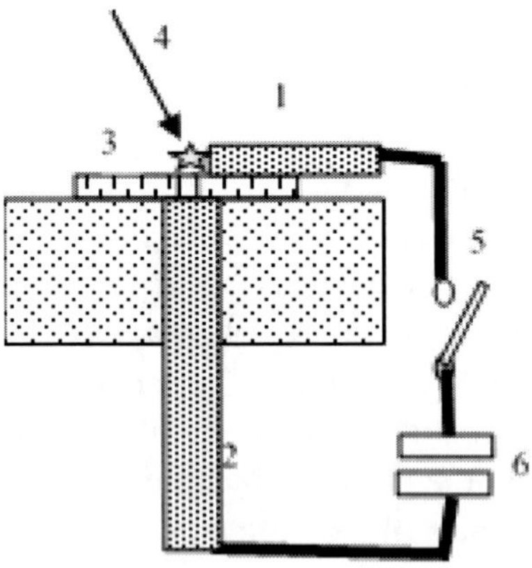

Figure 36. Production of AGO utilizing the erosive discharge. Here 1 and 2 are metallic electrodes, 3 is a plate for the erosion with hole 4. Capacitor 6 is discharged through switcher 5.

The erosive discharge promoted an occurrence of shells of the circulating light from the compressed air, vapors of metals, or other gases. A lifetime of the circulating light surpasses the lifetime of usual white light in a terrestrial atmosphere by approximately two orders of magnitude. It is explained by features of the molecular light scattering in the shells of circulating light. Unlike usual light scattering in three-dimensional space where the scattering light disappears in free space, a significant part of the scattering light in the shell remains in the shell and continues to circulate within it, but in other directions. It results in increasing the ball light lifetime.

The further increase in the lifetime of the circulating light can be obtained by a substantial growth of intensity of light in the discharge gap. Such physical conditions take place at strikes of usual natural linear lightning. The air pressure in the shell increases to such a degree that the molecules within the air become packed closely. It leads to a substantial growth of the circulating light lifetime because fluctuations of the air density that are responsible for the molecular light scattering in the shell disappear almost completely [7].

The following properties of AGOs were noticed. Their density is comparable with the density of atmospheric air; they have the low gas temperature, the small intensity of light radiation, and the high density of the stored energy. They are characterized by various interactions with various materials. For example, they can burn through a metal foil, but cannot penetrate through a usual sheet of paper. They aspire to keep their integrity at a meeting with obstacles. Their lifetime is abnormal great as compared with the lifetime of an ideal plasma. Authors mark a similarity in behaviors of AGO and Ball Lightning.

Let us consider an explanation of the selective action of the AGO on obstacles keeping in mind that the AGO and the ball light are the same. When ball light comes nearer to a sheet of paper, the regions of the paper located closest to ball light is heated up because the absorption by the paper of the light radiated from ball light (sometimes tracks of carbonization are seen on a paper). The paper heats a layer of air between ball light and paper owing to the phenomenon of heat conductivity. The reflective index of the hot air decreases. Since the circulating light moves along a gradient of a reflective index, the ball light jumps aside from the paper.

Coming nearer to a metal foil, the ball light also tries to heat the foil by radiation of light. But the heat conductivity of the foil is greater by 2 000 times than that of the paper, and the thermal capacity of the foil is greater by 15 times than that of the paper. As a result, unlike the paper, a significant heating of the foil does not occur. Heat spreads on the surface of the foil. It allows the

AGO to come nearer to the surface of the foil at a distance that is essentially smaller than the minimal distance at which AGO can come to the surface of the paper.

Finally, AGO also bounced off the foil. But a completely new mechanism can take effect sometimes at this time. At a sufficiently small distance from the surface of the foil, the AGO can heat a small portion of the foil to such an extent that the evaporation of metal begins. This leads to the appearance of metal vapor in the region between the AGO and the foil. The reflective index of the metal vapors is greater than that of the air. As a result, the gradient of the reflective index directed to the surface of the foil appears.

Moving along this gradient, AGO settles down in the maximum of the reflective index that is in immediate proximity from the foil surface. Note that the pressure of metal vapors is maximal on the foil surface as metal vapors move from the foil surface due to this pressure. However, the AGO moves along the gradient of the reflective index and is not blown up by the stream of metal vapors. On the contrary, the AGO moves in the direction opposite to the movement of the metal vapors to the area with the maximal reflective index. The area is located in the region from which the metal vapors move. The AGO settles down in the region where evaporation of the metal takes place and resides there all time until either its energy exhausts or it burns through the foil. In the first case, a characteristic crater is formed in the foil. Photos of similar craters are present in many publications. In the second case, a characteristic hole of a correct round form in the foil is formed. When the ball light has burned through a hole in the foil, it penetrated through the hole on the opposite side of the foil because the air on the opposite side is colder.

The same effect is observed when natural Ball Lightning collides with metal objects. For example, in case 8, Stakhanov describes a situation where Ball Lightning burns through a hole in a metal factory pipe. The Singer cites a photograph showing how Ball Lightning damaged the surface of a metal propeller in the place where the propeller creates maximum air pressure (see Figure 14).

One feature is observed when Ball Lightning interacts with metal objects. The Ball Lightning prefers to interact with objects stretched in only one direction. For example, Stakhanov describes a case where Ball Lightning flew into a room, went up to a carbine hanging on a wall, and vaporized a part of a metal ramrod. At the same time, there were no traces of reflow at the end. There are numerous reports when Ball Lightnings vaporized tens of meters of metal wire between the poles of an electric line.

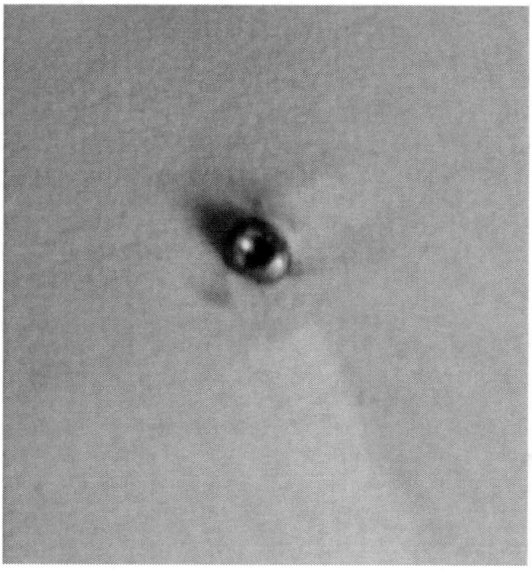

Figure 37. Spherical hollow in a tungsten rod.

Our explanation is as follows. By firing holes at the ends of an elongated metal body, the circulating light tells us that it can heat the metal only if the cooling of the metal due to thermal conductivity is small. In an elongated body, heat spreads only in one direction inward to the body. On the surface of an ordinary body, heat spreads in three directions (inward, sideways, up, and down). Thus, a degree of heating depends on a shape of the body and on the place where the heat source is located. The ends of the elongated body are heated to a higher temperature. If the heating temperature exceeds the evaporation temperature, then an erosion of the metal occurs.

Finally, the end face of a tungsten electrode is shown in Figure 37. A spherical hole is melted without any signs of fusion. The electrode was used to obtain the luminous silicon balls described above. Notc that the melting temperature of tungsten is 2700°C and the melting temperature of silicon, from which the balls are obtained, is only 1400°C.

How could this happen so that a molten drop of silicon in the gap between two electrodes could burn out a spherical hole in a tungsten electrode, the melting point of which is 1400 C higher than the temperature of the liquid with which this electrode is in contact? This liquid splashes out of the discharge gap and falls down in a form of several luminous drops onto various substrates. These drops turn into balls with anomalous properties caused by the light circulating over their surface. However, the light can circulate not only in the balls, but also in the drop itself in the inter discharge gap. The light is generated at the surface of the drop by an exothermic oxidation reaction of silicon, which leads to an increase in the temperature of the silicon oxide at the surface. As a result, the arising light is radiated in all directions. Part of the radiated light remains in the transparent spherical film and begins to circulate there. Gradually, the energy of the circulating light accumulates. The circulating light begins to heat nearby objects. If the temperature of the metal object exceeds the evaporation temperature, then metal vapors are drawn into the circulating light, since they have a refractive index greater than the refractive index of air. The circulating light heats the metal and draws in its vapors until its energy runs out. Thus, the same processes that are responsible for burning holes in metal foil are responsible for the formation of a hole in the electrode.

Circulating Light at Gas Discharges with the Evaporation of Metals

There are experiments where the evaporation of metal takes place directly at an electrical discharge. AGOs produced at the evaporation of metals were studied in St. Petersburg [94]. Such AGOs in the form of brightly shone yellow-red balls of 1 – 2 mm diameter appear at an erosive gas discharge with electrodes made of lead, tin, or aluminum. Their falling on a horizontal firm surface is accompanied by bouncing at the height of several centimeters. The number of jumps can reach up to 50.

The height of jumps gradually decreases, but the brightness of AGO luminescence does not change. At a final stage of the AGO movement, bouncing gradually passed in sliding, then in rotating and then they stopped. There were traces in the form of the dark circles surrounded by a light deposit on the surface of the table. The circles were located on those regions from which AGOs jumped aside. Traces of carbonization were seen on the surface of the wooden table in these places. The AGOs luminescence usually stopped suddenly at any point of its trajectory.

As is seen, the behavior of these AGO reminds that of the considered early luminous liquid drops of silicon. Traces in the form of dark circles surrounded by a light deposit on the surface of the table testifies that there is an exothermal chemical reaction that provides a long lifetime for these AGOs.

Carbonization of regions on the surface of the table from which the ball light jumped aside, confirms our assumption that coming nearer to the surface of the table, the ball light heats this surface. The sudden termination of a luminescence of the ball light, as well as the sudden disappearance of the ball light, was explained early.

Circulating Light in an Ultrasound Air Stream

It is interesting enough experimental data on the research of AGO behavior in a non-uniform atmosphere. A scheme of experimental installation is shown in Figure 38 [95]. From working chamber 1 of volume nearby two cubic meters, the air has been pumped out up to pressure P_0 = 0.02–0.04 atmosphere. After opening electromagnetic gate 4, the atmospheric air enters the chamber through a nozzle. There is supersonic stream 3 of 3,6 cm in diameter, moving at at a speed of about 470 m/s at the outlet of the nozzle.

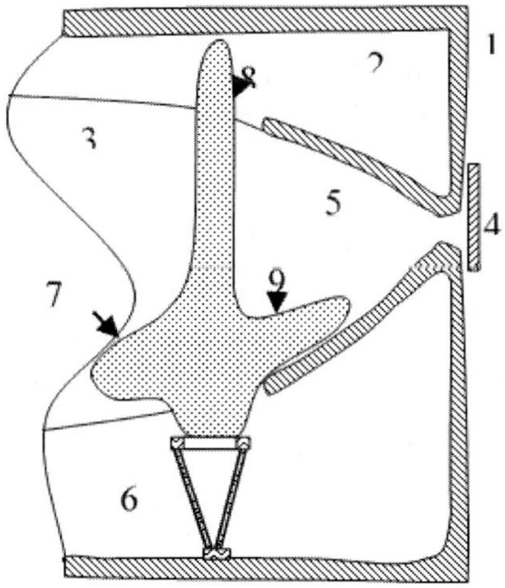

Figure 38. Schematic illustration of objects flown out plasmotron 6 in the supersonic stream from nozzle 5.

The discharge space of the AGO generator has a form of a truncated dielectric cone 6 with bases of 3 and 10 mm, respectively, and 30 mm height. The internal surface of a cone was covered with wax or stearin. One of the electrodes has been connected to the smaller basis; the second electrode in a form of a ring has been connected to the greater basis. The round aperture in the second electrode provided an output of AOs arising in the discharge space at the gas discharge. The speed of AGO in that area where it met the supersonic stream was about 50 m/s. As follows from a photo presented in [95], the products arising at the gas discharge, getting in a supersonic stream, were divided into 3 shone parts 7, 8, and 9.

Part 7 consists of the exited isolated atoms that move along the stream. As one would expect, this part has a bright white color. Authors mark “the existence of such strange thing, as spreading plasmas 9 along the border of a supersonic jet upwards in a direction to the nozzle." If to admit, that part 9 consists of ball lights, they should move along a gradient of a reflective index of that environment in which they are located. In the considered case, this direction coincides with a direction of the gradient of the air pressure in the supersonic stream. This gradient is directed to nozzle 5.

It is interesting to consider a physical nature of part 8. This part of 2–3 m long remains strictly collimated, has a bright purple color and a diameter of about 1 cm. Measurements show that the speed of sound in this part is about 1600 m / s. This is 5 times greater than the speed of sound in the air under normal conditions. The speed of sound in gases increases at gas is compression if the air pressure increases more than the density of the air. As shown below, if the density of air increases 1000 times, the air pressure increases 17000 times, and the speed of sound increases 17 times. In the case under consideration, the gas pressure is increased slightly less, since the speed of sound has increased by only 5 times.

Consider the mechanism responsible for such a great increase in the pressure in area 8. The electric current that passes through the discharge gap causes a sharp increase in the temperature in the gap. But a distribution of the temperature along a radius of the cone is non-uniform. Near to the lateral surface covered by wax, the temperature is minimal, as evaporation of the wax reduces the temperature. As the air pressure air at various points of the gap is identical, the reflective index n is maximal near the lateral surface of the discharge chamber.

Therefore, the ball lights appearing in the discharge gap move in a direction to the lateral surface of the discharge chamber. Having reached a lateral surface, the ball lights merge with other similar ball lights and form a great ball light, size of which gradually increases, until a diameter of its shell does not become equal to the diameter of the discharge chamber. The intensity of light in the shell constantly increases owing to a merge to the shell new ball lights. It leads to an increase in the air pressure inside the shell.

The following circumstance testifies that the shell receives light from small ball lights as a result of merges. Numerous ball lights of small diameter (some tens micrometers) have greater radiating losses. These losses depend on the wavelength of the light circulating in the shell. The greater wavelength, the greater the radiation losses. As a result, the light radiation of a long-wave range of the spectrum is radiated more quickly. As a result, such ball lights get a violet color. This phenomenon explains the violet color of the great ball light in part 8.

The air density and the air reflective index n in the discharge gap gradually decrease, as the hot air extends and leaves the discharge chamber. As a result, moving along a gradient n, the ball light leaves the discharge region too. The same concerns to that part of a shell of the great ball light which is located near the great basis. This part starts to penetrate the supersonic stream but does not move with it because a part of the shell of the great ball light, located near the small basis remains in the discharge chamber.

Notice, that a diameter of this part coincides with a diameter of the great basis. The part of the shell of the great ball light located near lateral walls of the discharge chamber continues to absorb small ball lights that continuously appear in the whole volume of the discharge gap. In this case, the great ball light proves as a unit with the shell where the intensive light circulates in every possible direction. Such ball light reminds the deformed Ball Lightning penetrating through a small hole. If the lifetime of such ball light would be great enough, its form could be spherical after the termination of the discharge.

Wax in this experiment plays a triple role. First, at the evaporation of the wax, there are gaseous hydrocarbons. Their reflective index is greater than that of the air. Such a mixture of the air and hydrocarbons is favorable for a generation of ball lights. Secondly, an evaporation of the wax provides an existence of great n for a long time in the discharge gap as new portions of the wax turn in vapors. It favorably affects the process of an accumulation of light in the great ball light. In the third, an evaporation of the wax reduces the temperature near lateral walls of the discharge chamber that promotes an increase in the reflective index in this area and arising the gradient of the reflective index directed to this area. Such gradient provides a movement of again arisen ball lights in the zone where these ball lights are merged into greater ball light. As a result, the light intensity of the great ball light gradually increases.

Circulation Light in Saturated Vapors of Nitrogen

Rather impressive experiments on the interaction between anomalous glowing objects (AGO) and liquid nitrogen were carried out by Klimov et al. [96]. A scheme of the setup is shown in Figure 39. The intense light for the production of AGO was produced by different means, for example, by the gas discharge over the surface of the liquid nitrogen, by the impulse plasmotron, by the powerful UV flash, by the flash of the powerful flash lamp, by power pulse laser. The largest AGO (diameter 10 mm) was obtained at the arc discharge directly into liquid nitrogen. An appearance of AGO was observed when using electrodes made of various materials (iron, graphite, copper, aluminum, brass).

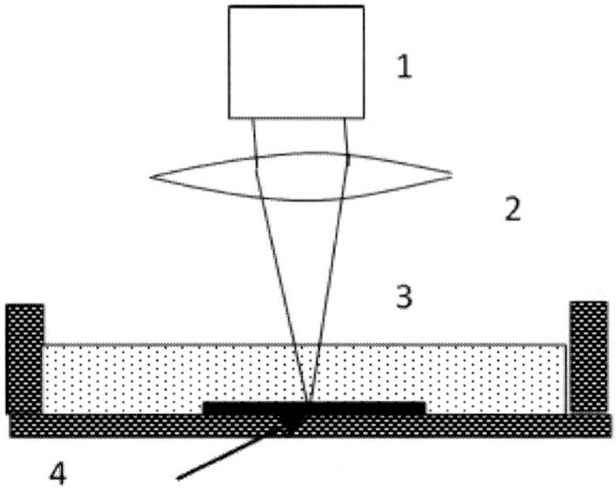

Figure 39. Production of AOs in the liquid nitrogen 3 utilizing powerful light beam from laser 1 focused by lens 2 on black paper 4.

Three types of long-living AGOs were observed during the experiment:

AGOs1 that are connected with the low intense luminescence evenly throughout the iquid nitrogen surfaceand the vapor above it.

AGOs2 in a form of bright spherical objects diameter of 5 cm;

AGOs3 in a form of dark spheroids with a diameter of 1 to 2 cm.

The lifetime of AGOs1 is about 10-20 seconds. The lifetime of AGOs2 is 20-40 seconds. AGO1 color is light blue. AGO2 color changed during the evolution from bright white to deep purple. AGO3 color is dirty grey. The energy stored in AGO2 could reach 100-200 J. AGO2, and AGO3 did not appear regularly in the experiment. Ought to note that the luminescence of the liquid nitrogen was observed not only at AGO production in gas discharges but also at AGO production utilizing the burst of a conventional powerful flash lamp used in a photo technique (the burst energy was about 600 J) [97].

The other most interesting results obtained in the experiments are the following:

1. Uniform blue luminescence excited in the liquid nitrogen was registered even when the UV lamp disposed near the surface of liquid nitrogen at one edge of the cell. The result is surprising since it is

assumed that the concentration of excited nitrogen molecules (and hence the luminosity of nitrogen) should decrease along with the distance due to the weakening of the intensity of UV radiation (proportional to R^{-2}).

2. Anomalous excitation of liquid nitrogen in the presence of various screens and diaphragms placed in the cuvette was investigated. In one experiment, a cuvette with liquid nitrogen was blocked by an opaque plate of 2 mm thick. Nitrogen was excited by a pulsed UV lamp only in one half of the cuvette. Then the partition was removed, but the sharp boundary between the excited luminous and dark unexcited parts of nitrogen was preserved. The sharp boundary remained even after forced stirring or shaking of liquid nitrogen with a dielectric spatula. If the partition had a round hole of 5 cm diameter, then in the second dark half of the liquid nitrogen a ball of excited nitrogen was observed, diameter of which was slightly larger than the diameter of the hole.
3. A transparent plexiglas plate of 2 mm thick was used as a partition. Liquid nitrogen in the second half of the cuvette was excited only in a rather narrow band near this partition. A shape of this luminous zone resembled a fringe. There was also a prolonged intense phosphorescence of the septum (even taken out of liquid nitrogen).
   A glow of the liquid nitrogen decreased sharply at placing a metal object in the cuvette or brought near thereto the solenoid being energized by an alternating current (frequency is 10 kHz, the strength of the magnetic field is 7957.75 A/m).
4. The target in the form of the black carbon paper was placed on the bottom of the cuvette for excitation of liquid nitrogen by laser spark as is shown in Figure 40. The target was placed in the focus of the lens. In control experiments without nitrogen, the torch was observed on the target, and the target was burning. The diameter of the burned hole did not exceed 5 mm. In the case where the ditch was filled with liquid nitrogen, a luminous spherical object of 10-20 mm diameter near the focal point of the lens was formed. The object was accompanied by several small secondary objects not larger than 5 mm in diameter arranged randomly on the bottom of the cuvette.
   AGO moved in a usual atmosphere to the cuvette with liquid nitrogen located nearby. After the termination of the gas discharge, the luminescence of the volume of the liquid nitrogen within 5 seconds was observed. The AGO with precise borders of 0.5-4 mm diameter,

which brightness surpasses a background luminescence of liquid nitrogen, can be seen at the bottom of the cuvette, on its internal lateral walls, and also on the surface of the liquid nitrogen. The spectrum of AGO radiation was in a range from 400 up to 500 nanometers. Radiation lasts during 10-30 s with gradually fading.

5. The cylindrical reflector made of aluminum or brass foil was placed around the cylindrical UV lamp as is shown in Figure 40. The lamp was placed in one of two focuses of the reflector. Light beams radiated by the lamp is concentrated in the second focus. The system was submerged in liquid nitrogen. The spontaneous appearance of bright white AGO2 of millimeter size near the second focus was observed at the moment when the lamp flashed. Local centers of melting or even burning reflector material were detected after 5-10 flashes of the lamp. These centers looked like craters of small diameter (about 4-5 mm) with black soot coating at their edges and the molten core of 1 mm diameter. The number of craters reached 5-7. No burning was observed in control experiments without nitrogen. The energy of the single AGO2 was estimated at 6 J.

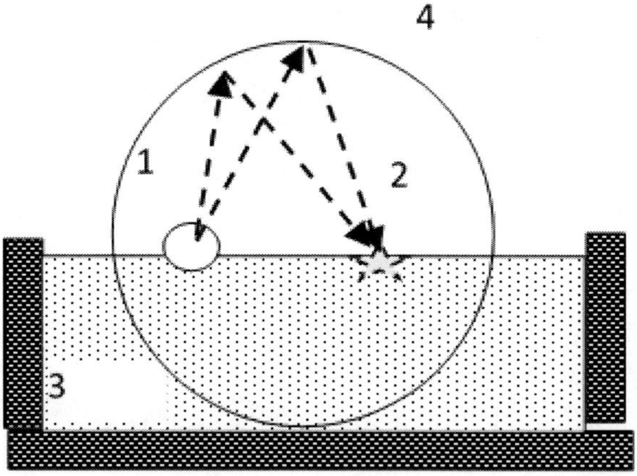

Figure 40. Production of AGOs 2 in the liquid nitrogen 3 utilizing UV flash-lamp 1 located in the cylindrical reflector 4.

Arising from the gas discharge, AGOs move in the normal air atmosphere into a nearby cuvette with liquid nitrogen. After termination of the gas discharge, the luminescence in the volume of liquid nitrogen is observed for 5 seconds. AGO with clear boundaries diameter of 0.5-4 mm, which exceeds the brightness of the background luminescence of the liquid nitrogen, can be seen at the bottom of the cuvette, on its inner sidewalls, and the surface of liquid nitrogen. The radiation spectrum is in the range from 400 to 500 nm. The radiation lasts for 10-30 seconds, gradually fading. The color of AGOs changes in the process of evolution from bright white to deep purple.

AGO in the liquid nitrogen may be prepared through the intense ultraviolet radiation. The flash of the powerful flash lamp (power of the capacitor bank 300 - 600 J.) also can be used for the production of identical AGOs. The time of AGO luminescence obtained by this method is 10 - 20 seconds. A similar picture was observed in the absence of liquid nitrogen when the water vapor was introduced into the gas discharge. However, the lifetime of the AGOs has been less than one second.

The fact that the radiation gradually fades indicates that the AGO shell is liquid. If the shell is composed of highly compressed air, AGO would cease to exist suddenly with a characteristic sound. This phenomenon is explained in [96].

Thus, after a seemingly primitive experiments, numerous anomalous phenomena were observed. Concerning circulating light, the following characteristic phenomena can be noted:

1. All experiments with the anomalous glow of liquid nitrogen in a cuvette, including partial penetration of the glow through the partition, as well as the glow of the partition removed from liquid nitrogen. These effects are explained by the presence of many small AGOs that appear after the flash of light.
2. The formation of craters on a metal foil. For AGO with circulating light, an appearance of craters on the metal foil is characteristic due to the very high energy density of the circulating light.
3. The appearance of luminous balls, the lifetime of which was several seconds, and the spectrum of their radiation gradually changed, shifting toward the long-wavelength side.

It is much more difficult to explain how the circulating light enters the droplet. In all previous cases, the circulating light appeared in the shell due to the light emission of excited atoms located directly in the shell. Such radiation gradually accumulated in the shell. In the case under consideration, there are no excited atoms. The light in the shell can be excited by external light only, but in this case, according to the theory, the excitation time is equal to the lifetime of the light in the shell. But the lifetime of the circulating light is measured in seconds, and the excitation time is measured in milliseconds. Nature offers the next riddle.

The solution is this. It is known that the lifetime of circulating light in an ideal spherical shell is determined by the inevitable radiation losses, which, according to the theory, depend, among other things, on a ratio of the length of the sphere's equator to the wavelength of the circulating light. For droplets of liquid nitrogen, the losses sharply increase when this ratio becomes less than 10. In this case, the lifetime of the circulating light is measured in tens of light wave periods. The lifetime is much shorter than the duration of the excitation. Therefore, a steady-state of circulating light is established in such a drop, but its intensity differs insignificantly from the intensity of the external light. As the size of the liquid nitrogen droplet grows due to a condensation on its surface of molecules of the cold oxygen surrounding the drop, the losses of circulating light in the droplet decreases, and the ratio of the indicated intensities increases. Ultimately, the droplet reaches a size where the radiation losses becomes negligible and the circulating light accumulated during growth can exist for a long time [48].

Circulating Light at Discharges in Ionized Air

Another rather indicative experiment on studying the properties of AGOs is connected with the analysis of reflection from a flat wall of the shock sound wave propagating in small-ionized air [98]. A scheme of experimental installation is shown in Figure 41. Here 1 is a working section of a shock pipe of 10x10 cm^2 cross-section in which the shock wave 2 at 400-600 m/s speed of a triangular pressure behind its front propagates. In the working section of the shock pipe, the decaying pulse discharge with an average density of a current 30 mA/$см^2$ and 3 ms duration of burning was created. The discharge was created directly ahead of the arrival of the shock wave.

Conventional air of 0.003-0.03 atmospheres pressure was used as a working gas. The sizes of the plasma area were made 10x10x20 cm^3. In the plasma zone at a distance of 15 cm from its beginning, the flat quartz wall 7 was settled down for reflection of the shock wave. The air pressure was measured using the piezoelectric sensor 6 located at 5 cm distance from the quartz wall. As shown in [98], the air pressures fixed by the sensor behind the falling and reflected waves decreases in the presence of the gas discharge.

For an explanation of this abnormal phenomenon, it is possible to admit that at this time, there are numerous small ball lights. The pressure inside the ball light shell is much greater than that in the surrounding space. As a result, the average pressure of the remaining air fixed by the sensor decreases. Like previous cases, small ball lights can merge into one great ball light.

This assumption is supported by information, that "during experiments occurrence of specific plasma formation with unusual parameters near the wall in the region of the hot layer was revealed." This formation occurred after the reflection of a shock wave from wall 2. Initially, this formation covered an area of 3.5 cm, but then it was gradually compressed up to 2 cm."

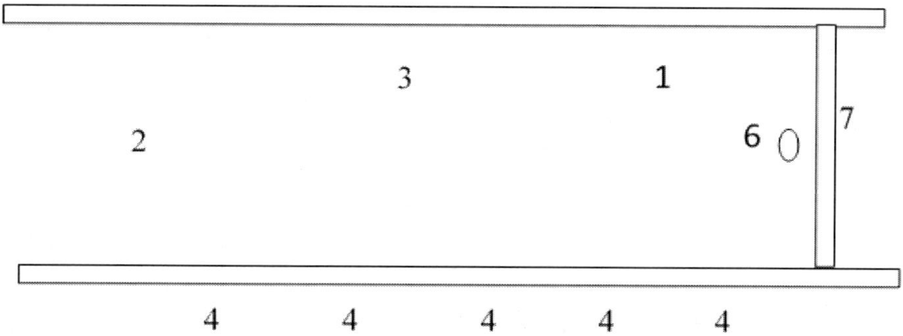

Figure 41. Scheme of setup for measuring parameters of shock sound wave traveling through a glowing discharge.

Circulating Light at Discharges on a Surface of the Water

Egorov's and Stephanou's experiments concerned the production of long-living luminous objects in a glass of water [99] has been repeated in many laboratories around the world. In these experiments, the role of wax plays a drop of water placed on one of the electrodes, as shown in Figure 42. The gas discharge occurs between electrodes 2 and 3. Electrode 2 represents a copper ring placed on a bottom of a polyethene vessel of 18 cm diameter. The vessel is filled with usual water. A height of the water column is 18 cm.

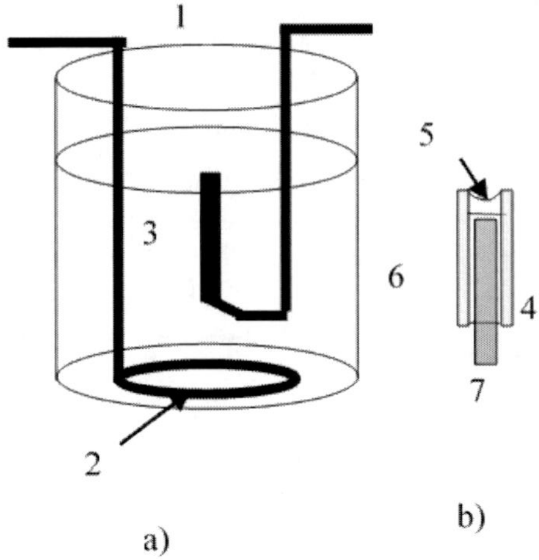

Figure 42. Production of a luminous formation in a vessel with water.
a) location of electrodes in water;
b) structure of central electrode 3 where quartz tube 4 with a metal or coal electrode 7 of 5-6 mm in diameter contains about 0.1 milliliters of water 5. The tube towers on the height up to 5 mm above a water level 6 in the vessel.

A structure of electrode 3 is shown in Figure 42b. Here the quartz tube 4 with a metal or coal electrode 7 of 5-6 mm diameter contains about 0.1 milliliters of water 5. The tube towers on a height from 3 up to 5 mm above water level 6 in the vessel. The light gas in the form of vapor between the electrodes was formed upon the water evaporation induced by the electric discharge of the capacitor battery that had a capacitance of 600 μF and was charged up to 5 kV. Egorov and Stephanou emphasize that there was an optimum potential difference between electrodes. This is likely associated with the water evaporation regime. Water is sprayed at high voltage.

The discharge was accompanied by the appearance of a luminous cloud above the vessel, which slowly rose upwards. The lifetime of the cloud was about second. The evidence that the film exists and the light circulates is the sharp boundaries of the film, which are formed by the light inside it. In this case, there is no closed spherical film. However, the light introduced into such a cloud cannot be radiated into free space, since it experiences total internal reflection at the film-air boundary.

Circulating Light Penetrating through a Transparent Wall

From the presented consideration can be concluded, that the gas discharge in a hermetic vessel is favorable for the generation of ball light. Really, in this case, the density of gas in a vessel and, hence, its reflective index can increase only at the gas discharge because of evaporation of electrodes. Experimental confirmation of this assumption is presented in [89]. A scheme of installation in which the AGO is generated is shown in Figure 43.

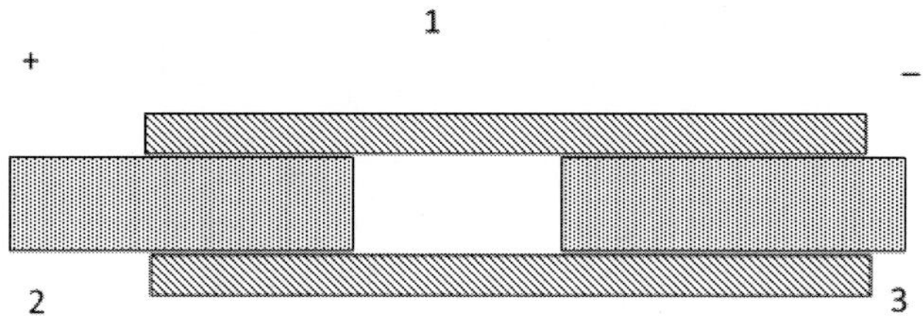

Figure 43. Scheme of discharge camera between electrodes 2 and 3 where autonomous objects penetrating through the transparent wall of tube 1 have been produced.

To provide full tightness of the internal chamber around electrodes, the bandage has been placed. Video recording of the process shows that after the termination of the discharge in a tube, it got a barrel-shaped form. Inside the tube, a weak luminescence was seen that suddenly disappeared. At this moment some shone AGOs in diameter about 1 mm appeared outside the tube. Their occurrence was accompanied by crackling. On the several frames of the video, it is visible, that the AGOs float in the air, and then fall. AGOs that fell on a surface of a table jumped several times and disappeared completely. The tube remained tight.

Meaning, that the walls of the tube are impenetrable for any particles (a molecule, atoms, ions, electrons), it is necessary to recognize, that all such particles remain inside the tube. As for light, it passes through the transparent walls of the tube almost free. The process of penetration of ball light through transparent walls of a tube is similar to the process of penetration of Ball Lightning through windowpanes.

The fact the ball lights left the tube with some delay can be explained as follows. At the gas discharge, there is an erosion of the electrodes, and pressure in the tube increases. The swelling of a tube testifies to it. The reflective index in the tube at the great pressure is greater than the reflective index in the space outside the tube. In this case, the ball lights remain inside the tube. After the termination of the gas discharge vapors are condensed, and pressure in the tube falls. With a reduction of the pressure with cooling electrodes, the reflective index decreases and can become smaller than that outside the tube, as the temperature inside the tube is greater than outside. At that moment, the ball lights penetrate outside through the transparent walls of the tube.

A similar situation is described by Stakhanow in event number 110. The luminous ball of 2-3 mm diameter had flown from the signal neon ball light at the moment when the ball light was turned on. The ball light is burned out. This experiment was repeated successively once more. The luminous ball of 3-4 mm penetrated through the glass balloon of the ball light at the moment when the ball light was burned out. The ball fell on Batman sheet paper, jumped several times on it at the height of several centimeters, and disappeared without leaving a trace. However, attempts to repeat this result in the further experiment were unsuccessful.

The study of the spectral composition of the radiation of autonomous objects was carried out by Barry, who confirmed that the radiation of autonomous objects is the emission of excited gases whose reflective index exceeds that of air and nitrogen. Radiation of carbon dioxide, oxide of nitrogen, ozone, and methane were detected. Besides, metal lines were discovered from which the electrodes were made.

The analysis of the spectral composition of the radiation of autonomous objects made it possible to draw several conclusions about the conditions of the formation and behavior of autonomous objects. These additional conclusions became possible under the assumption that autonomous objects contain circulating light.

First of all, nature made it clear that an increase in the reflective index of air in the shell of the ball light can occur not only by increasing its density due to the compression but simply by pulling gas molecules from the surrounding space in which the reflective index is greater than the reflective index of air. For example, the reflective index of carbon dioxide is $n = 1 + 5.15\ 10^{-4}$, and the reflective index of air is $n = 1 + 2.78\ 10^{-4}$. Replacing air molecules with carbon dioxide molecules in the shell is much less energy-consuming than compressing air by almost 2 times to get the same increase in the reflective index.

Circulating Light at Vacuum Discharges

There are strong reasons to believe that features of vacuum discharges can be explained by an existence of the ball lights. As is known, vacuum discharges can be accompanied by the occurrence of avalanches of electrons, each of which contains 10^9-10^{11} electrons. A similar avalanche has been called ecton by Mesyats [100, 101]. Similar vacuum objects were considered by K. Shoulders [102]. He calls them Exotic Vacuum Objects (EVO). The nature of the appearance of such objects is not explained, but there is the agreement that they have anomalous properties.

As shown in the section Circulating light during gas discharges with the evaporation of metals, AGO has the property of vaporizing the metal upon contact with it. In this case, the evaporation occurs until the energy of AGO is exhausted.

We can assume that an avalanche of electrons is generated at the cathode during the evaporation of the metal. As is known, the cathode is heated in the radio tubes so that electrons can fly out of it. The phenomenon of secondary electron emission at the anode is also known, in which electrons are knocked out of the anode. To eliminate this effect, radio tubes were invented with an additional grid, called tetrodes.

We try to explain the appearance of the AGO at a vacuum discharge as follows. First, what is the cause of the electron avalanche? Earlier, we considered the case when AGO heats the metal and leaves a crater in it. It can be assumed that an avalanche of electrons occurs since AGO heats the cathode. The avalanche of electrons arises from a locally heated cathode due to the conventional thermal emission of electrons. The limited number of electrons in the avalanche is since the AGO energy is finite. Thus, each AGO produces one avalanche of electrons.

The evaporation of the metal and heating of the cathode are accompanied by the usual phenomenon of the thermos emission used in radio tubes. When the energy of AGO becomes smaller than a threshold, AGO disappears. As a result, being near the surface of the cathode, the AGO causes the issue of an avalanche of electrons. We should notice that the characters of the craters formed by AGO on metal surfaces at erosive gas discharges [89] and the craters formed on the surface of the cathode at vacuum discharges [100] are identical.

An avalanche of electrons moves towards the anode, and the energy of each electron increases by several hundred volts. Upon collision with the anode, metal vapors arise, from which AGO is formed. It can be assumed that there is an erosive discharge at the anode, where metal at the anode undergoes erosion. In this case, such an AGO can be positively charged, since electrons are absorbed by the positively charged anode. As a result, such an AGO moved in the direction opposite to the movement of electrons, that is, in the direction towards the cathode. Such an AGO was investigated by Shoulders, and he calls it EVO. Approaching the cathode, the AGO forms the next avalanche, and the process repeats.

The description of the casual movement of the AGO in the Plante experiments reminds the description of the casual movement of the cathode spot at vacuum discharges [101]. Probably, the AGO is responsible not only for the occurrence of ectons but also for the occurrence of cathode spots.

Until now, a mystery that arose more than a century ago has not been solved. The movement of the cathode spot in a magnetic field does not correspond to theory. In our opinion, the movement of the cathode spot coincides with the movement of the AGO. When approaching the cathode, the AGO chooses the region of the cathode where the reflective index of the metal vapor near the cathode is maximum. With other things being equal, this takes place in the region whose temperature is minimum. This leads to the fact that the position of the next cathode spot is shifted relative to the previous one.

Thus, the energy of the circulating light accumulated in AGO arising at vacuum discharges is anomalous great and is comparable with the AGOs energy arising at other erosive discharges.

Circulating Light in Cavitation Bubbles

With sharp mixing, cavitation bubbles appear in the liquid. Surprisingly, such bubbles can glow. For example, the glow of cavitation bubbles can be observed at night from the stern of the ship, where you can see the faint blue glow of a jet of water foamed by the ship's propeller. The word "cavitation" is derived from the word "cavity." Empty bubbles are formed inside the liquid and become spherical under an action of the surface tension forces. Under the same forces, the cavitation bubbles gradually decrease in diameter and eventually disappear.

A regular study of properties of such bubbles began after it became possible to artificially create such a bubble in a fixed place in the liquid. For this, a standing spherical acoustic wave is set up within liquid. The bubble is formed in the minimym of the pressure in the standing wave. This wave in the rarefaction phase creates a very low pressure in the water, which leads to a local rupture of the water and the formation of one cavitation bubble. During about a quarter of the ultrasonic wave period (that is, while the pressure remains very low), the bubble grows, and if the standing sound wave is spherically symmetric, then the bubble remains spherical. In some experiments, the bubble diameter reached fractions of a millimeter.

In the compression phase, the cavitation bubble collapses faster and faster. The collapse process is also accelerated by the surface tension force.

In the final fractions of the period, a very short and bright flash of light escapes from the center of the vanishing bubble. Since the cavitation bubble is born and disappears with the frequency of the acoustic wave, the bubble emits an averaged sonoluminescent light. This phenomenon is called sonoluminescence. The word "sono" indicates that the phenomenon is associated with sound, but the phenomenon is more associated with radiation of light and manifests itself in the form of the glow of cavitation bubbles.

Such light has several paradoxical properties or, in our terminology, has anomalies. Some of them are associated with the conditions for the occurrence of such light. For example, a small concentration of inert gases dissolved in water greatly enhances the effect. The brightness of the light increases dramatically when the water is cooled. The light that flashes from the bubbles lasts between 35 and a few hundred picoseconds long, with peak intensities of the order of 1–10 mW. When the bubbles emit light, they are about 1 micrometer in diameter. The size depends on the ambient fluid (e.g., water) and the gas content of the bubble (e.g., atmospheric air). The addition of a small amount of noble gas (such as argon, neon, or xenon) to the gas in the bubble increases the intensity of the emitted light.

In the greatest degree, the following anomalous phenomenon surprises and discourages scientists. A bright flash of such light has, as a rule, a more or less smooth spectrum, without any separate spectral lines. This spectrum grows steeply towards the violet side and is approximately similar to the spectrum of radiation from an absolutely black body with the temperature of the order of hundreds of thousands of Kelvin. Detection of very high bubble temperatures by spectral methods is limited due to the opacity of liquids to the short-wavelength light.

It was the spectrum that became the main stumbling block when trying to explain the phenomenon. If such light is of thermal origin, then it is necessary to explain how ultrasound heats cold water to such temperatures. As is known, the water evaporates at a temperature of 100 degrees Celsius. Why doesn't water evaporate at temperatures a thousand times higher than 100 degrees Celsius?

If the high temperatures have nothing to do with it, then what is the general origin of light? It is unclear how light, the spectrum of which corresponds to the radiation of a black body heated to a temperature higher than the temperature of the sun, can appear in cold water.

There is every reason to believe that such anomalies were not without taking into account the circulating light. The suspicion of the circulating light in the bubble is heightened when one considers that such bubbles have a destructive effect on metal objects (for example, a ship's propeller). In the first chapter, we gave an example in Figure 14 of the destruction of an aircraft propeller blade by the circulating light in a form of natural Ball Lightning. We also mentioned that the resulting AGOs could burn through or leave a crater in a thin metal foil. A long-term repeated exposure of such AGOs to the foil will lead to an appearance of many craters, that is, AGOs will have a destructive effect on the foil. In the next chapter, we will try to explain the role of circulating light in the occurrence of this major anomaly.

In contrast to the cavitation bubble that occurs during acoustic cavitation, which is located in a fixed place and occurs periodically, there is the hydrodynamic cavitation that produces many cavitation bubbles in the entire area where the physical conditions are favorable for their occurrence. These bubbles move along with the fluid flow. When observing bubbles through the transparent glass, the presence of cavitation bubbles appears as a cloudy area in transparent water. It is not possible to trace the evolution of these bubbles. In addition to the appearance of a cloudy area, cavitation bubbles manifest themselves in the fact that over time, corrosion of objects occurs in the area where cavitation bubbles appear.

There is an explanation for the physical processes in which a cavitation bubble can cause damage to objects, in particular metal ones. The essence of the explanation boils down to the fact that when a cavitation bubble collapses, it releases a jet under high pressure, which damages the object. However, no calculations confirming this hypothesis are presented. There are also no recommendations to ensure an absence of the corrosion during cavitation. Recommendations are reduced to a list of measures that allow you to avoid cavitation. The measures are associated with a decrease in the speed of water flows, which limits a performance of pumps, pipelines, and the speed of ships.

We present our explanation using the concept of the circulating light, keeping in mind that cavitation bubbles obtained by another method when studying the phenomenon of sonoluminescence contain the circulating light. This fact makes it possible to explain all the observed anomalies of such cavitation bubbles [26].

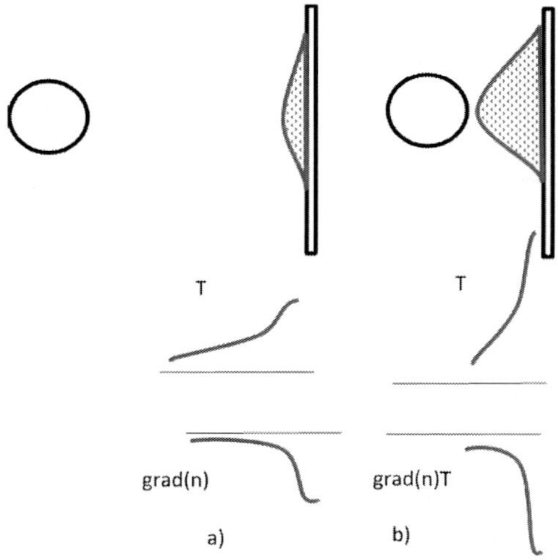

Figure 44. Distribution of the temperature T and the gradient of the air reflective index grad(n) near a vertical paper sheet at approaching the circulating light. a) great distance; b) small distance.

In addition, the presence of the circulating light in luminous objects with anomalous properties, which are studied in attempts to obtain Ball Lightnings in a laboratory, has been shown [25]. The anomaly of these objects is that, they bounce off a vertically placed sheet of paper in their horizontal movement, but burn through or leave craters on a thin metal foil. An explanation of this anomaly is given in [25], where it is shown that a behavior of such luminous objects is explained by the peculiarities of the behavior of the circulating light near objects with different thermal conductivities, heat capacitiesy, and conditions for propagation of heat during their local heating.

When an object with the circulating light approaches a sheet of paper, the paper is heated in places closest to the object due to the light emitted by the luminous object (Figure 44a). An area with the increasing temperature heats the nearby layers of air due to the phenomenon of the heat conduction. In this case, the closer the air is to the plane of the paper, the higher its temperature, as shown in the graph in Figure 44b. As a result, the refractive index of the heated air decreases and the gradient of the refractive index directed away from the sheet of paper arises. The distribution of the negative gradient of the refractive index is shown in the same graph. Once in this area, the circulating light begins to move along the gradient of the refractive index, that is, it bounces off the sheet of paper.

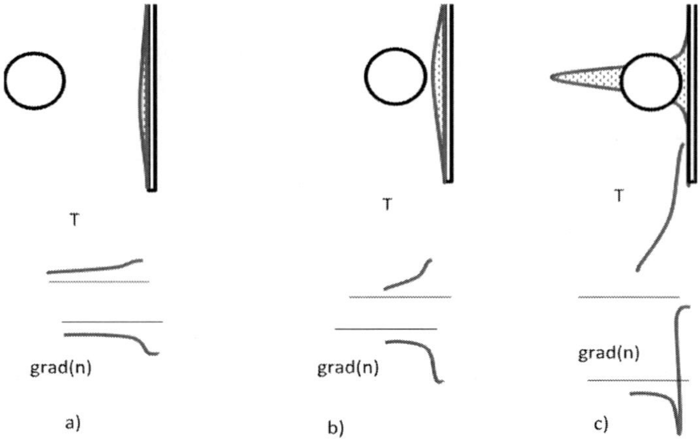

Figure 45. Distribution of the temperature T and the gradient of the air reflective index grad(n) near a vertical metal foil at approaching the circulating light.
a) great distance;
b) intermediate distance;
c) small distance where the evaporation of the metal vapors takes place.

A different situation occurs when the circulating light interacts with a thin metal foil. Coming nearer to the metal foil, the circulating light also tries to heat up the foil by radiation of light. But the heat conductivity of the foil is greater by 2,000 times than that of the paper and the thermal capacity of the foil is greater by 15 times than that of the paper. As a result, unlike the paper, significant heating the foil does not occur. Due to the high thermal conductivity, heat spreads in all directions, which leads to heating a larger area, but with the lower temperature, as shown in Figure 45a. In this case, the circulating light can approach the foil surface at a much smaller distance, as shown in Figure 45b. In this case, two situations are possible. First, the circulating light can bounce off the foil, as it does with the paper. Secondly, the temperature in a local area in the immediate vicinity of the circulating light may rise to such an extent that the metal begins to evaporate from the heated local area. In this case, the distribution of the refractive index gradient in the immediate vicinity of the metal surface changes fundamentally.

Since the refractive index of metal vapor is much greater than the air refractive index and the concentration of metal vapors is maximum at the very surface of the foil, then there is a positive gradient of the refractive index directed towards the surface of the foil, as shown in Figure 45c. Moving along this gradient, the circulating light continues to evaporate the metal from the local area until it burns a through-hole in the foil, or leaves a crater in the local area if the energy stored in the circulating light is not enough to burn the foil through. As was shown earlier, the circulating light draws gases from the surrounding space into its shell, the refractive index of which is greater than the refractive index of air [23]. The same phenomenon explains a favorable effect of an erosive gas discharge on the formation of anomalous luminous objects [24].

The cavitation bubbles with circulating light that arises in water behave in exactly the same way as the considered luminous objects with the circulating light when interacting with metal in the air.

From the presented mechanism of an appearance of craters on metal surfaces, it follows that the metal can be protected from cavitation corrosion by applying a protective coating to the metal surface. The coating should have a minimum thermal conductivity and heat capacity. The thickness of the coating should be greater than the depth of the craters, that is, should be greater than the diameter of the cavitation bubbles, the size of which is several tens micrometers.

There are objects with unique properties in the world of circulating light. Taking into account the circulating light makes it possible to explain many natural phenomena with anomalous properties. In particular, the explanation of the physical nature of corrosion occurring during cavitation allows to undertake attempts to exclude this phenomenon, which will lead to a positive effect in many areas of modern technology, in particular, it can increase a performance of pumps, pipelines and the speed of river and sea vessels.

Circulating Light Appearing in Rooms from Sockets and Nails

In the world of circulating light, there are not only objects that were obtained when trying to get Ball Lightning in the laboratory but also objects that have a relation to circulating light but were created by nature itself. Below we list cases when objects with the circulating light appeared directly in rooms in front of the amazed audience. These objects are closer in their parameters to natural Ball Lightning than to objects obtained in the laboratory. First of all, they surpass the latter in size and duration of existence. A distinctive feature of such objects is longtime of their occurrence. Below we will give some cases of observation of such objects to have an idea of what is spoken about.

There are several reports that Ball Lightning appears directly on the premises from loudspeakers, telephones, electric meters, electric lamps, and simply from sockets. It is usually noted that initially Ball Lightning is similar to foam, that is, consists of many luminous bubbles. Then a single luminous ball can be formed from them. We give a few observations of such objects taken from [2]. We save message numbers so that the reader can read them in the original if desired.

No. 72. Ball Lightning of 10 cm diameter jumped out of the body of a telephone set hanging on the wall, and the lid fell off. This happened after a discharge of a linear lightning. The ball rolled across the floor and exploded loudly, awakening several sleeping people, but not harming any of them.

No. 76. At the same time as a strong lightning discharge took place, a crack appeared near the radio receiver into which the reproducer plug was not inserted. A dark red ball hung at a distance of less than a meter from the outlet. After some hesitation, the ball floated in the direction of the open door, passed less than a meter from the eyewitness (and he did not feel the heat), went out into the corridor, and then through the second open door to the street. There it exploded loudly, leaving no consequences at the place of the explosion. The external input of the radio receiver was damaged and burned out, the outlet melted.

No. 77. Once at 12 o'clock, a thunderstorm began; there were nine people in the workshop. Suddenly there was a sharp hiss from the loudspeaker, so everyone thought that the radio was starting to broadcast. However, instead, a yellow stream about 10 cm long began to flow out of the socket terminal into which the loudspeaker plug was inserted, which quickly twisted into a ball. On the surface of the ball, irregularities similar to spikes formed. The ball glowed in orange, and the spikes were slightly darker. Within 2-3 seconds, the hissing ceased, and a luminous spherical mass was formed, from which a slight rustling emanated. The ball went down to a height of 1 m, smoothly made a circle around the room, circling a 3 m long work desk in the middle of the shoe workshop. Returning to the socket, the ball fell into a corner and disappeared with an explosion and a bright flash. A spot of charred paint of 25-30 cm diameter remained on the floor, soot on the stone wall. The loudspeaker has not deteriorated.

No. 78. After the discharge of linear lightning, which broke several pillars of the radio transmission line, a sphere with a diameter of 20 cm appeared from the outlet. It went to the stove and disappeared into it, merging with the edge of the ajar door. The outlet was also slightly damaged.

№ 79. There was a thunderstorm. Ball Lightning appeared in the room. After a strong crackle in a socket for the radio speaker, a fiery cylindrical cord about 45 cm long and 10 cm diameter broke out. Then the tail of the cord broke off from the outlet, and the cord turned into a regular ball with a diameter of about 15 cm. The ball glowed like a 10-20 W lamp with a yellowish light with a blue tint. It slowly sailed about 1.5 m at the distance of 3 meters from the observer, then it stopped, hung in the air, and disappeared silently. All this lasted about 5 s. The ball hung above the bed at the height of 30 cm from the blanket, on which, however, there were no traces. After the end of the thunderstorm, the radio was turned on. The radio network worked properly

No. 112. The author saw how a luminous orange ball jumped out of an electrical outlet when her six-year-old son tried to stick two copper wires there. The ball was 25-30 cm in diameter. Passing by his son and bypassing his mother, it moved to a height of 1.5 m from the floor to the open window, near which it disappeared with the sound of a shot. All this lasted 10-12 s.

No. 113. The author plugged an electric capacitor into a conventional power outlet (220 V). A strong crack was heard, and along with sparks, a formation appeared from the outlet about 10 cm in size, a pear-shaped shape that emitted white-blue light. It was only 5 cm from the fingers of the author. Rising about 15 cm, Ball Lightning began to fall on the floor at the speed of a child air ball. A faint crack was heard. Having fallen to the floor from a height of about 1 m, it began to move slowly, jumping, bouncing from the floor to the radiator. Walking about a meter, it disappeared. In those places where it touched the floor, there were burns from paint with a diameter of 1-3 mm.

No. 114. The author writes that during a short circuit in his apartment, there was a luminous ball of 10-15 cm in diameter. The place where the short circuit occurred was invisible because of a furniture wall. There was a clap and a bluish ball of 10-15 cm diameter appeared from the gap about 1 mm wide between the furniture and the floor. It rolled on the floor about a meter and went out, leaving no consequences. The room was dark. The current in the circuit was less than 6A since the fuses were not knocked out. The ball had a regular shape and a slightly blurred border. The intensity of the light is like that of a 0.5 W flashlight. The observation duration is about I s. The insulation was burnt after a short circuit at the contact point.

No. 81. There was a roll of thunder. Climbing to the second floor of his apartment, the boy heard a hiss from behind. Turning around, he saw a luminous ball slowly appear, gradually increasing its size. The ball emerged from a protruding nail near the window. Having reached a diameter of 10-12 cm, the ball came off the nail and began to move after the people climbing the stairs. They hid in apartments. Father tried to close the door behind him. However, the ball managed to slip into a narrow gap - less than 1 cm wide. Having penetrated through the door, the ball flew at a distance of 0.5 m from the boy, blowing it with a stream of slightly warm wind, and disappeared behind the toilet door. They no longer saw the ball. All this lasted 40-50 s. Later, a hole 2 mm in diameter was discovered in the glass of the toilet room. No flashing detected.

No. 82 Making sure that rain was not expected shortly, the author drew attention to the sound (weak buzzing) coming from the top of the column, which stood on a site near the house at a distance of about 20-25 meters. The column was old, and the wiring was already removed from it. Looking closely, he saw a red spot the size of a cherry, appearing on a metal hook in the upper part of the column, on which an insulator was attached. The hook was at the height of 4 m from the surface of the Earth. The luminous ball began to grow, changing color from red to yellow, and then to green. At the same time, the luminous flux from it grew, so that if earlier it could be compared with a lamp of several watts, now it glowed like a 500 W lamp. All this took about 2 seconds. Having reached a size of 15-20 cm in diameter, the ball broke away from the column and went, slightly deviating to the sides, first to the observer, and then gradually turning to the tree, which grew next to the neighboring two-story house 5-7 m from the column. The ball moved at a speed of 0.5 to 1.5 m / s. Having travelled about 10 m in 10 s, the ball came across a tree and exploded with a strong sound.

No. 84. The author saw from the balcony how, during a severe thunderstorm, outbreaks of "flame" appeared near the horizontal metal pipe located on two vertical pillars. Then a bright white ball with a diameter of 15-20 cm slid from the end of the metal pipe. The ball went down and began to move horizontally at the height of about 1 m from the ground. The ball disappeared around the corner of the house. After 30-40 seconds, a strong crack was heard.

115 Author was fixing up a TV. When he brought the screwdriver to the picture tube to check for the presence of high voltage, a spark began to slip through. There was a crack in the kitchen. Upon reappearance, his daughter, whom he asked to see what was happening, said that lightning from the switch had slipped in the kitchen. The day, however, was cloudless, hot and there were no signs of a thunderstorm. After seven or eight repetitions, the discharges ceased, but there was a crack near the place of entry of the wiring and the meter. Turning around, he saw that an orange ball of 10-20 cm diameter of regular shape with a sharp border was hanging near the entrance, which was luminous like a 50-100 W lamp. After 10-20 s, the ball went out. It turned out that there was a short circuit.

The simplest assumption is the following. It is known that the space around the Earth is immersed in an electric field, the magnitude of which varies depending on the place and changes in time. It is assumed that this field is created mainly by charged clouds. In this case, the space between the cloud and the Earth can be considered as a great capacitor, the energy of which is comparable to the energy of linear lightning that is about $E = 10^9$ J. The first plate is a charged cloud. The second plate is the ground. Firstly, we can see that the energy of the capacitor is greater by 4-5 orders of magnitude than the energy stored in Ball Lightning that is estimated in the range 10^3- 10^5 J.

On the assumption that the area of the cloud is 1 km^2 and the distance to the ground is 1 km, we have that the energy density W in this cubic kilometer is about 1 J/m^3. In this case the strength of the electrical field is about E= 0.5 MV/m. It is smaller than threshold E_T=2 MV/m that is required for the air breakdown.

The situation changes if there are bodies in the form of conductors or dielectrics in this field. Under the influence of an electric field, the uniform distribution of electrons in the body is disrupted. As a result, a non-zero surface charge density arises, which determines the strength of the field oriented perpendicular to the surface of the conductor. In a dielectric, the polarization of the dielectric takes place. As a result, the dielectric is also unevenly charged. It is known that the field strength created by charges in a conductor or dielectric can significantly exceed the field in which they are located. This takes place near the surface of great curvature. This field may exceed the threshold E_T. As a result, a breakdown occurs, and a current appears. The electrostatic distribution of charges disappears.

This phenomenon is not new and is observed in the form of St. Elmo's fires as well as the corona discharges at high voltage electrical lines. There is plasma in the region as well as molecules of ozone and nitric oxide. The excited atoms of nitrogen and oxygen fluoresce with blue or violet light. The same mechanism causes neon signs to glow.

Charging a wooden object with arising a corona discharge is not surprising. For example, corona discharges in the form of St. Elmo's fire are observed on masts of a ship. Corona discharges can be seen on the high voltage electrical lines that are isolated from the ground. Additional high voltage can be supplied due to the high strength of the electrical field produced by a charged cloud. Additional measures are undertaken in the form of additional wires located above the main wires. The additional wires are connected with the ground on each mast to which wires are suspended.

Electrostatics disappears with corona discharge but the conventional picture with electrical current, voltages, resistance, the Kirchhoff laws does not appear. First of all, this concerns the absence of closed wire circuits where currents flow. A corona discharge occurs near one electrode. The second electrode is located at an indefinite distance, and the current path to it is also indefinite. This favors the occurrence of Ball Lightning for two reasons.

First, in contrast to the previously considered methods for producing anomalous glowing objects (AGO), in which AGO occurred in the discharge gap between two electrodes, the distance between which is limited, the absence of the second electrode at the corona discharge allows one to obtain an AGO of a much larger size.

Second, the nature of the corona discharge does not change with increasing discharge power, since the second electrode is removed by hundreds of meters at least. If the second electrode is located at a relatively small distance from the first, then with an increase in the voltage between them, a nature of the discharge changes. Tht corona discharge with a relatively small discharge current passes into a spark, the current of which increases by orders of magnitude. In this case, a continuous discharge passes into a pulsed one, resembling the discharge of ordinary linear lightning. As all observers note, the time of occurrence of Ball Lightning from metal objects is several seconds, at least. Therefore, a spark discharge is not involved in the formation of Ball Lightnings.

The physics of the corona discharge is well studied. Ionization of molecules is produced not only using the electrical breakdown of air but also electron avalanches. The strength of the electric field decreases and the formation of plasma ceases with moving away from the sources of the corona. The region where plasma is formed is seen as St. Elmo's fire. However, a cloud of electrons exists in the space between the first and second electrodes. The charge of this cloud depends on the distance to the second electrode and its size.

Thus, there are all the prerequisites for the occurrence of circulating light. There is plasma with the reduced reflective index, which can confine the circulating light, there are molecules of ozone and nitric oxide with the increased reflective index, which can be drawn into the shell of the circulating light, thereby increasing its reflective index and decreasing its radiation losses.

The electrons that are produced in the plasma move in the air and ultimately reach the positive pole (cloud). The positively charged ions are neutralized by electrons moved upward from the ground. As a result, a continuous current arises between cloud and ground that discharges the capacitor formed by cloud and ground. The current value is determined by the voltage between cloud and ground and the resistance between them. A value of this resistance is determined by the electrons formed at the corona discharge. This resistance is extremely great. As a result, the current in the circuit is constant. That is why the luminous of St. Elmo's fires is constant in time.

A similar situation occurs when an autonomous object is created in continuous mode. For example, this is a discharge between wet surfaces or between jets of water. In this case, the current density in the discharge does not increase due to the limited power of the high voltage source. The same blue and violet glow is seen in Figure 28 in the area where the plasma is formed.

Where can a powerful source be taken? The conventional corona discharge power is insufficient. Besides, corona discharges are observed much more frequently than an appearance of Ball Lightning and a duration of their glow exceeds orders of magnitude a duration of the formation of Ball Lightning, measured in seconds. We need to show how natural Ball Lightning can be formed from a rare situation of several seconds duration.

Let us first try to consider the process of the occurrence of Ball Lightning from a metal hook driven into a wooden pole presented in case number 85 [2]. The hook is charged initially due to small currents in the form of electrons that left the hook due to the same phenomena that are responsible for corona discharges in the form of St. Elmo's fires on wooden masts of a ship. The hook is charged by an extremely small current because electrons must pass a long distance about 1 km toward the cloud.

Let us now assume that the situation is changed in a short time and a negatively charged cloud appears above the hook and, therefore, a direction of the vector of the strength of the electrical field is changed. In this case, the capacitor formed by the hook and ground is discharged along the path of 10 m length. The resistance of the path decreases by 100 times and the discharge current increases accordingly. Besides, electrons begin to move in the opposite direction toward the hook and further toward the ground. The total charge of these electrons can be great enough.

At the same time, the resistance between the hook and ground is still great enough to prevent the instantaneous discharge that many researchers used to obtain Ball Lightnings in a laboratory. The charge is accumulated for several seconds. This time is enough to accumulate light from a luminous plasma.

Perfectly different forming Ball Lightning occurs when Ball Lightning arises from sockets and reproducers. In this case, the functions of the hook are performed by the long wires connected to the outlets. In rare cases, the strength of the atmospheric electrical field on one end of the wire can be significantly greater than that on another. On the other hand, it is well known that the voltages of any electrical conductor in all its places are the same. Otherwise, electric current will flow from a place where the voltage is greater to a place where the voltage is less until the voltages will be the same.

The wire can be quite long and the atmospheric voltage at the place where one end of the wire is located can be quite different from the atmospheric voltage where the other end of the wire is located. However, the voltage across the entire surface of the wire is the same. How can this be? Very simple. The charges in the wire are distributed in such a way that the voltage from these charges compensates for the difference in atmospheric voltages at the ends of the wire.

Now suppose that the end of the wire opposite the socket is connected to ground through some kind of resistance. The situation under consideration is equivalent to that in which the considered wire is in a vertical position in the electric field of the Earth, directed also vertically. In this case, the redistribution of charges inside the wire also occurs, and charges with a high density appear at the upper end of the wire. A corona discharge occurs in which electrons flow from the upper end of the wire, and the wire is replenished with charges coming from the ground. In fact, there is a slow discharge of the capacitor formed by the cloud and the ground. The horizontal wire can be quite long, which contributes to a higher charge density at its end. At the same time, for the formation of such a density, a rather rare condition must be satisfied that the atmospheric voltage at one end of the wire significantly differs from that at the other end.

As a result, we have a free high-voltage source with a large internal resistance, which does not allow large currents to flow through itself, since in this case the main voltage drop will be on this large resistance. The energy to power this source is taken from atmospheric electricity. A similar home-made source based on a transformer for a microwave oven is shown in Figure 27 and has been used to produce objects with circulating light near wet surfaces and metals.

A situation is possible when the charge density on the wire end is sufficient to produce the corona discharge. Analog of the corona discharge is seen in Figure 28 in a form of violet column. The energy of the corona discharge is taken from the electrical field of the capacitor between a cloud and ground due to a slow discharge of this capacitor through the following elements: cloud, air between cloud and wire, wire, corona discharge, ground. As a result of the power corona discharge, a set of small balls arises that are attracted to each other like soap bubbles in soap suds. Unlike soap bubbles, the light bubbles tend to merge because the larger bubble of light takes the circulating light away from the smaller one.

When trying to reproduce the described process in the laboratory, it should be borne in mind that the Ball Lightning, the stored energy of which is at least several kilojoules, should be formed within a few seconds. The power supply must have a large internal resistance so that the discharge current is limited by this resistance. Therefore, the power of the power supply must be at least several tens kilowatts.

Based on the foregoing, it can be assumed that the analogue of Ball Lightning from the outlet can be obtained using a powerful high-voltage power supply, the internal resistance of which can be regulated. Taking into account the current advances in electronics and electrical engineering, a source can be designed in which the voltage and internal resistance can be regulated, including software.

Our general considerations are confirmed by the results of attempts to obtain Ball Lightnings with a powerful continuous discharge [113]. Authors underline that a powerful continuous high-voltage power supply is required. As such, a Tesla transformer was used, capable of delivering continuous power of 3.2 kW. According to Tesla's recommendation, rubber-coated copper wire or a charcoal pin were used as an electrode. It was noted that a lit wax candle favorably influenced the formation of luminous beads.

It was noted that luminous balls several millimeters in size appear near the electrode. Some balls seem transparent; through them, other streamers are visible. It was observed how the initial 6 mm ball grew into a fiery red globule 5 cm in size within 1 second. Ultimately, all balls instantly disappear. The authors note that the power of their installation is approximately 100 times less than the installation of Tesla, which received and observed long-lived Ball Lightning.

We can only notice that the natural installation can even exceed the Tesla installation in both power and voltage. We can conclude that full-fledged Ball Lightnings can be obtained by using high-voltage power sources, which in a few seconds can produce energy of hundreds of thousands of joules. The engines of many cars have this capability.

This is an initial attempt to consider the occurrence of Ball Lightning. Other attempts were made above where the process of the occurrence of Ball Lightnings and Bead Lightnings from linear lightnings is considered based on the presented laws and phenomena inherent to the circulating light [11].

Conclusion

We described only a small part of experiments in which, in our opinion, anomalous properties of observed luminous objects are associated with the circulating light. Numerous experiments with electric welding in water, experiments with vacuum discharge, experiments with electrolysis in heavy water were left aside.

We can conclude that there are absolutely new objects and phenomena in nature, about an existence of which nobody earlier knew and suspected even. These are ball lights of various kinds which size, lifetime, and stored energy can be changed over a wide range. A shell of such ball lights can be the most various also. It can consist of compressed air, a mixture of gases which reflective index is greater than that of the air, vapors of metals. The ball light can be in a form of a transparent liquid droplet where the intense light is circulating along its surface in all possible directions. Ball light can be in the form of molten material, wherein the exothermic chemical reaction occurs with the ambient air on its surface. The reaction provides the long luminescence of the ball light for about ten seconds.

The most convincing, in our opinion, are experiments with jumping luminous balls, the movement of which is anomalous in several aspects.

First, the jumps of such balls cannot be explained by laws of mechanics even if all known types of forces are taken into account. A simple frame-by-frame analysis of the frequency of jumps shows that this frequency is more than 2 times smaller than the bounce frequency of ordinary solid-state balls at an elastic surface. Attempts to explain this phenomenon by the fact that the force of gravity is partially compensated by the buoyancy force of Archimedes leads to the fact that the number and height of bounces should decrease sharply, which contradicts the video.

Further, none of the sane people can imagine that a ball can bounce off more than two dozen times on a simple sheet of A4 paper lying on the horizontal surface of a conventional table.

Finally, another mystery is that at the end of a conventional 3 mm diameter tungsten electrode with a melting point of about 2700 ° C, a spherical hole can be melted, although the silicon is liquidated at 1400 ° C.

We have shown that these anomalies can be explained by an assumption that an action of the circulating light is taken into account.

Studying anomalous phenomena in which circulating light can take part, we have more than once been able to make sure that nature offers the most unexpected options that scientists could not think of.

This was the case with an increase in the reflective index of the shell in which light circulates. We have long studied physical phenomena that could lead to an increase in the air reflective index in the shell due to the light circulating in the shell. We even managed to discover a new phenomenon, according to which an increase in the reflective index of air in the shell is possible only if the light intensity exceeds a certain threshold. Nature has proposed a simpler way to increase the reflective index in the shell where intense light circulates. If ball light is located in a mixture of gases, an increase in the reflective index of the shell is carried out by drawing into the shell the mixture components, the reflective index of which is greater than the air reflective index.

So it was with the appearance of ball lights in an atmosphere of saturated vapor, where the initial formation of large fluctuations is possible. Quite unexpectedly, experiments showed that the formation of ball lights above liquid nitrogen in a field of intense light is possible.

Thus, nature prompted us the first significant refinement of our relative pure model of circulating light. This refinement leads to several facts that were not initially taken into account.

1. We believed that the gradient of the air reflective index of the atmosphere is determined by two parameters, temperature, and pressure. Now we must introduce the third parameter. This is an inhomogeneity of the distribution of impurities in the air. For example, in the space where smoke leaves the chimney, it must be borne in mind that the closer to the chimney, the greater the concentration of carbon dioxide and the higher the reflective index in this region.
2. When analyzing methods for producing autonomous objects, we often came across the term erosive gas discharge. In this case, the metal electrodes are coated with various substances, in most cases, hydrocarbons, for example, wax. It was noticed that this coating favorably affects the production of autonomous objects. This phenomenon is explained as follows. During the discharge, the temperature in the interelectrode gap increases due to the electric current. At high temperatures, the decomposition and evaporation of hydrocarbons occur. As a result, the reflective index increases. This favorably affects the time of formation of the circulating light and, consequently, the time of its life.

Besides, the evaporation of hydrocarbons requires energy, which contributes to a decrease in temperature in the interelectrode gap, which contributes to an increase in the reflective index.

3. Another technique that was widely used by experimenters to improve methods for producing circulating light was to partially seal the discharge chamber. For example, the discharge occurs in the glass tube, and communication with the outside world is through a small hole at the end of the tube [3]. Sealing the discharge chamber ensures that when the temperature increases in it, the reflective index does not decrease, since this increases the air pressure and the air density remains unchanged.
4. The fact that the shell of circulating light contains pairs of metals has allowed us to explain a strange at first glance phenomenon, according to which emerging autonomous objects manifests itself as follows. In a collision with a thin metal foil, they burn holes in it or leave a crater in it. However, this effect is absent in a collision with a normal sheet of paper.

The only anomaly that we could not explain is the indifference of the scientific community to unknown natural phenomena that can be reproduced in any garage and studied in detail. Nobel laureate Russian scientist Piotr Kapitsa said that science and technological progress should be different. Science is called upon to generate new knowledge like a miner extracts coal, and technological progress is to introduce new technologies based on new knowledge. There are many scientific organizations in the world whose names contain the word "science." However, after the invention of lasers in 1960, it is difficult to list new scientific discoveries. They can only appear when studying phenomena that science has not yet been able to explain.

Chapter 3

Circulating Light in a Micro World

> Nature is our kindest friend and best critic in experimental science if we only allow her intimations to fall unbiased on our minds.
>
> M. Faraday

Introduction

We planned a book of two chapters. In the first chapter, we describe the nature of Ball Lightning in the form of a spherical shell of strongly compressed air in which ordinary white light circulates in all possible directions and prove that this is indeed the case, believing that natural Ball Lightning, the properties of which are described and documented, is the best experimental sample, indicating that all properties of the circulating light, which directly follow from the generally accepted laws of optics and physics, fully correspond to the properties of experimental samples produced by nature.

In the second chapter, we intented to show that Ball Lightning is not the only object in which the circulating light is responsible for its anomalous properties. There are a number of anomalous objects obtained in quite different ways, in which the circulating light is responsible for their anomalies. Thus, we have shown that a special world of circulating light exists in nature and naively believed that the study of its unusual properties should attract the attention of theorists and experimenters. However, our expectations were not met.

Convinced that there is a whole world of circulating light in nature with its own unique properties, it would be foolish not to analyze how these properties can be used. While skeptics treat this phenomenon with distrust, we have a unique opportunity to use this circumstance to analyze known anomalous phenomena in order to find out whether circulating light is involved in these anomalies.

Since 2003, we have found quite a few anomalous phenomena in which circulating light is responsible for their anomalies. As already mentioned, we have shown that the circulating light is responsible for the apparently extremely high temperature in the luminous ball resulting from sonoluminescence. The observed shift of the measured spectrum to the short-wavelength region is explained not by the high temperature of the ball, but by the compression of the circulating light with a decrease in the ball diameter. Thus, we have shown that attempts to use the phenomenon of sonoluminescence to carry out nuclear reactions are futile.

We have also shown that circulating light in cavitation bubbles causes corrosion of metal objects during cavitation, and proposed means to eliminate the harmful effects of cavitation on corrosion.

Bearing in mind the phenomenon according to which circulating light draws into its shell the components of a gas mixture with the highest refractive index, one could use these phenomena to separate the components of a gas mixture consisting of different isotopes of the same element, when chemical methods of separation are powerless. According to Table 1, the refractive index increases with increasing molecular weight. Therefore, it can be assumed that isotopes with a higher atomic weight may have a higher refractive index.

However, a more important question remains open. Is it possible to develop new technologies and devices using the unique physical parameters of objects with circulating light? First of all, this concerns ultrahigh air pressure in a spherical shell, at which the air density increases by more than 1000 times and reaches the density of water. In addition, the energy density of the circulating light exceeds the energy density of light propagating in a straight line by billions of times. In this case, the optically induced forces increase billions of times and exceed other types of forces in many cases. This is responsible for an anomalous behavior of circulating light in the Earth's atmosphere. These properties of circulating light also appear in other situations. Situations are known in which circulating light demonstrates its high energy density. This is evidenced by observed cases when circulating light burns through metal sheets and leaves craters on metal objects.

Studying the above objects with anomalous properties from the world of the circulating light, we drew attention to the fact that many researchers of these objects indicate that the appearance of these objects is accompanied by excess heat. This is especially true for objects that arise during cavitation. At present, so-called vortex cavitation heat generators are manufactured and sold. They produce heat energy more than consume electricity and violate the Joule law and the law of conservation of energy. Other researchers note that during erosive electrical discharges, elements are transmuted and their isotopic composition changes. This can only happen if nuclear energy is involved in these processes.

At present, questions about the possibility of extracting energy from nuclei are quite well studied. From the point of view of modern science, two processes for obtaining nuclear energy are known. These are Fission and Fusion. The first process occurs in atomic bombs and nuclear power plants. The second process takes place in the hydrogen bomb. For more than 70 years, the first process has been successfully used in nuclear power plants. For more than 60 years, unsuccessful attempts have been made to use the second process for the production of electricity.

The third process is known, called Cold Fusion. It received this name because, unlike the classical Fusion, for which a temperature of about hundreds of millions of degrees is required, that process occurs at much lower ordinary temperatures. In 1989, two electrochemists, Martin Fleischmann and Stanley Pons, reported that their apparatus produced anomalous heat ("excess heat"), the magnitude of which, they claim, defies explanation except in terms of nuclear processes. Many scientists have tried to replicate the experiment with some of the details available. Hopes faded with the appearance of a large number of negative repetitions, the withdrawal of many registered positive repetitions, the discovery of shortcomings and sources of experimental errors in the original experiment. By the end of 1989, cold fusion claims were considered dead by most scientists, and cold fusion subsequently gained a reputation as a pathological science. In 1989, the US Department of Energy (DOE) concluded that the published results of excess heat did not provide conclusive evidence for a useful source of energy and decided not to allocate funding specifically for cold fusion.

The excess energy and transmutation of nuclei are mentioned not only in the context of Cold Fusion, but also in many other experiments with objects where the circulating light presents. For example, the annual Russian conference is called "Ball Lightning and Cold Transmutation of Nuclei." Because of this, it is worthwhile to analyze more carefully what connection is possible between the circulating light and the excess energy. At the same time, we believe that there is nowhere for excess energy to come from, except from nuclear energy. Thus, there is some kind of connection between circulating light and nuclear transformations.

The first step is to identify the difficulties in carrying out nuclear transformations and how circulating light can help overcome these difficulties.

As is known, Fusion processes require fuel and a confined environment with sufficient temperature, pressure, and confinement time to create a plasma in which fusion can occur. The combination of these conditions is known as the Lawson criterion. In stars, the most common fuel is hydrogen, and gravity provides extremely long confinement times that reach the conditions needed for fusion energy production. Proposed fusion reactors generally use heavy hydrogen isotopes such as deuterium and tritium (and especially a mixture of the two), which react more easily than the most common hydrogen isotope, to allow them to reach the Lawson criterion requirements with less extreme conditions. Most designs aim to heat their fuel to around 100 million degrees, which presents a major challenge in producing a successful design.

The strong force acts only over short distances (at most one femto meter that is equal to size of one proton or neutron), while the repulsive electrostatic force between nuclei acts over longer distances. In order to undergo fusion, the positive charged fuel atoms need to be given enough kinetic energy to approach each other closely enough to overcome the electrostatic repulsion. The amount of kinetic energy needed to bring the fuel atoms close enough is known as the Coulomb barrier. Ways of providing this energy include speeding up atoms in a particle accelerator, or heating them to high temperatures.

It is not difficult to verify that two out of the three listed Lawson conditions are satisfied in the shell of circulating light. Indeed, there is an extremely high pressure and the time of possible interaction between neighboring nuclei is even longer than necessary. Only high temperature is missing. The question arises. Is it possible to do without high temperature? According to the generally accepted opinion, this is impossible. Otherwise, such a nuclear transformation with the release of excess heat would have been carried out.

However, there are numerous reports of fixation of excess heat. There are various theories, but they are not generally accepted. The main drawback of these theories is that they do not indicate ways to improve devices that generate excess heat.

Keeping in mind the fact that there is a high pressure of the gas in object from the world of the circulating light, we will try to develop a theory of nuclear reactions in which the main condition is the presence of high pressure. As well as the development of the theory of Ball Lightning, this is a problem with a known answer in advance. Thus, it is required to develop a theory, from which it follows that high pressure is required for the implementation of nuclear reactions.

Since the mention of excess energy occurs quite often in the context of anomalous luminous objects, the influence of circulating light on the occurrence of excess energy should be considered. Circulating light is a common phenomenon for various anomalous objects, so it is natural to assume that it can somehow influence the occurrence of excess heat.

Are Nuclear Reactions Possible without Overcoming the Coulomb Barrier?

Before proceeding to further consideration, it is necessary to resolve a fundamental issue. Is it possible to noticeably transfer energy from one object to another if the force of interaction between them is extremely weak? Indeed, a distance between nuclei is greater by a factor four orders of magnitude than a diameter of nucleus. For example, if we image a size of nuclei is 1 m then a distance between nucleus of such dimensions is equal to 10 km.

The answer to this question is no. But there is a particular case the answer can be yes. The following conditions should be performed. First, the force applied to the object should not be constant, but change periodically over a long period of time with the same frequency. Secondly, the object must be an oscillatory system in which the energy of oscillations is stored for a long time. Thirdly, the natural frequency of the oscillatory system must coincide with the frequency of the external periodic force. Fourth, the phases of the weak periodic force must be favorable for energy transfer to the object.

An example of such an oscillatory system is an ordinary children's swing, consisting of two poles and a horizontal crossbar, to which two ropes are attached. A seat is attached to the rope ends, on which the child sits. Another child can swing this swing, applying a small force to each swing at that moment and in such a way as to increase the energy of the oscillatory system, that is, the amplitude of the swings. Swinging takes time. The smaller the external force, the more time is required. It is possible that the swing will be swung to such an extent that it will break.

The question is where does a weak external force come from to shake the oscillatory system in the form of a nucleus? The source of such force can be another nucleus, and exactly the same nucleus. It can be assumed that the oscillation frequencies of these two identical oscillation nuclei are the same.

What are the grounds for believing that the nucleus is an oscillatory system? It follows from the most general considerations. A nucleus consists of nucleons that have mass. In addition, there are forces between nucleons that prevent the repulsive forces that occur between like charges. This is enough for the nucleons to oscillate. We do not know what frequency of oscillations, but we can assume that nucleons in identical nuclei oscillate with the same frequency.

There is one more question. What grounds are there for believing that there is something far from the nucleus that acts on another nucleus located at some distance? The answer to this question requires a separate consideration.

First of all, we should find out what is common between objects from the world of the circulating light and objects from the micro world.

Circulating Light and Wave-Particle Duality

The concept of wave-particle duality has been introduced into science since it has been experimentally established that corpuscles in the form of electrons, neutrons, etc. have wave properties. At the same time, we have shown that light pulses behave like corpuscles in the optical medium. They have mass, and their motion is completely subject to Newton's laws [74].

In this context, the circulating light can serve as an illustration of wave-particle duality. It consists of corpuscles in the form of oxygen and nitrogen molecules and waves in the form of electromagnetic waves with wavelengths ranging from 1.0 to 0.4 μm.

How much closer is the circulating light to corpuscles or waves? If we compare their masses, then the circulating light is closer to the corpuscles. The ball light has mass and volume and its velocity can be determined. The mass of ball light consists mainly of a mass of highly compressed air. The mass of the the circulating light can be neglected. In this respect, Ball Lightning is a corpuscle.

If we talk about wave properties that the stored energy in ball light significantly exceeds the energy of compressed air. In this respect, ball light is an electromagnetic wave. Moreover, the wave in a form of the circulating light is subject to an additional type of forces in a form of optically induced forces. In an inhomogeneous optical medium, these forces act on the circulating light and significantly exceed the types of forces acting on the particle in a form of compressed air.

If we are talking about the lifetime of ball light, then its lifetime is limited due to the inevitable radiation losses, at which part of the energy of the circulating wave is radiated into the surrounding space, and the stored energy in the ball light gradually decreases. This leads to an arising instability and ball light ceases to exist in a relatively short time.

If we are talking about the lifetime of a particle in the micro-world, then it would seem that its lifetime is unlimited. However, a closer analysis shows that the lifetime is limited. Indeed, a particle consists of nuclei that are subject to natural radioactive decay. The lifetime of nuclei can be equal to billions of years. However, this difference is qualitative. It can be argued that a particle and ball light have wave properties that lead to their decay ultimately.

The physical nature, properties, and behavior of the ball light are known, and the nature of the occurrence of the wave-particle dualism associated with circulating light is also known, since this phenomenon obeys the generally accepted laws of physics and optics. It is shown that the behavior of the circulating light is determined by its mass and the energy of the light stored in it [74]. Thus, we have a complete understanding of the properties of the circulating light, both in the form of a corpuscle and in the form of a wave.

It is interesting to compare a behavior of the circulating light in different situations with a behavior of objects of micro-world to find out possible similarities between the ball light and these objects.

Features of the passage of the circulating light near obstacles are considered in [34]. It is shown why the circulating light bypasses obstacles. This phenomenon is explained by the fact that the radiation radiated by the circulating light heats the surface of the obstacle, which, due to the phenomenon of heat conduction, heats the nearby layers of air. The closer the layer of air is located to the surface of the obstacle, the higher its temperature. On the other hand, it is known that the air density decreases with increasing the temperature. Since the refractive index is an increasing function of the air density, the refractive index of the air decreases as the heated surface of the obstacle is approached, that is, the gradient of the refractive index is directed away from the surface.

Thus, the optically induced force acts on the circulating light, the direction of which coincides with the direction of the gradient of the refractive index, that is, in the case under consideration, the force is directed away from the surface. Under the influence of this force, the circulating light is deflected from the obstacle. The deflection is accompanied by a decrease of the gradient of the refractive index and the circulating light continues its movement past the obstacle.

A similar situation occurs if the circulating light passes through a slit-like obstacle as shown in Figure 46. In this case, the circulating light heats both walls of the slit. If the circulating light passes exactly through the center of the slit, then both walls are heated equally and the circulating light is not deflected. If the trajectory of the circulating light is located closer to the right wall of the slit, then the right wall heats up more than the left one. The circulating light is repelled from the right wall more strongly than from the left. As a result, the circulating light is deflected to the left.

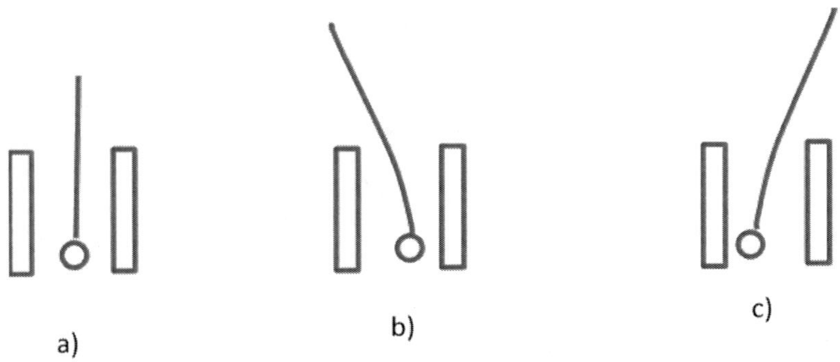

Figure 46. Trajectories of the circulating light at passing a slit where it is located
a) at the center of the slit,
b) near to the right wall,
c) near the left wall.

The value of this deflection depends on the gradient of the refractive index, which in turn depends on the wall temperature, which in turn depends on the heating time of this wall. This time is inversely proportional to the speed v of the circulating light. The circulating light experiences an acceleration under the action of an optically induced force F that is proportional to the energy of the circulating light and inversely proportional to the mass m of the compressed air in which the light circulates. Thus, the deflection of the circulating light that passes through a slit-like obstacle is in inverse proportion with a product of the mass and velocity.

On the other hand, the Bohr hypothesis about wave properties of corpuscle is based on experimental date that deflection of a corpuscle is also in inverse proportion with a product of the mass and velocity.

Thus, we have the experimentally established phenomenon, in which the particles of the micro-world, passing through the split, experience the influence of the split and change the trajectory of their movement. The phenomenon was called wave particle duality. It indicates that outside a particle of the micro-world there is something that, interacting with the walls of the split, affects the particle.

Like the circulating light, we can imagine the corpuscle that is not limited by the volume where its mass is concentrated. Similar to the emission of circulating light in the surrounding space, which ultimately leads to decreasing the energy stored in the circulating light, it can be assumed that a similar situation also takes place for the corpuscle. There is something outside the corpuscle that allows it to be influenced by the slit wall without touching it. It is necessary to assume that when passing by an obstacle, the corpuscle interacts remotely with the obstacle, that is, outside the volume in which the mass of the corpuscle is concentrated, something additional takes place that affects the movement of the corpuscle. This "something" we can clarify by considering the lifetime of circulating light and corpuscles.

We believe that there is one more evidence that something exists outside a nucleus that can influence on identical nuclei. This something is considered in the following section.

Disappearance of the Circulating Light and Decay of Nucleus

The question of the presence of some substance outside the volume in which the mass of the corpuscle is concentrated also arises when considering the features of the disappearance of the circulating light and the corpuscle. Knowing the physical nature of the circulating light, it is easy to determine the cause and features of its disappearance [37]. The lifetime of the Ball Lightning can vary from a few seconds to several minutes. It is shown that this time is determined by the inevitable energy losses due to light radiation into the surrounding space due to molecular light scattering on fluctuations of the air density. In this case, the intensity of the circulating light gradually decreases, which entails a decrease in the refractive index of the compressed air and a further increase in radiation losses.

Of course, there are no losses in the corpuscle due to the molecular light scattering. But for light circulating in a limited volume, there is another type of loss associated with the inevitable radiation of part of the circulating light into the surrounding space. This applies not only to an electromagnetic wave but also to any other types of waves propagating along a curved trajectory.

Indeed, the field of a light wave propagating along a 1D rectilinear light guide exits in free space and decays exponentially there. This field is called evanescent field. The field cannot excite a propagating wave in free space, which carries the wave energy from the waveguide into free space. This is explained by the fact that the spatial periods along the waveguide's axis and the spatial periods of any wave propagating in free space are different for any direction of the wave in free space.

A different situation occurs for a curved waveguide. In this case, as one moves away from the waveguide to the outer side, the spatial period of the exponentially decaying field gradually increases as is shown in Figure 47. As can be seen from the figure, the spatial period of the evanescent field increases with distance from the circulating wave. At some distance from the waveguide, this period will become equal to the wavelength propagating in the surrounding space. In this case, a wave is excited in the surrounding space by that weakened field that has penetrated through the tunnel between the waveguide's boundary and the boundary beyond which the wave is excited in free space. Thus, a bent waveguide always has energy losses, called radiation. Circulating light propagating along a closed path always propagates in a curved waveguide and, therefore, always has radiation losses.

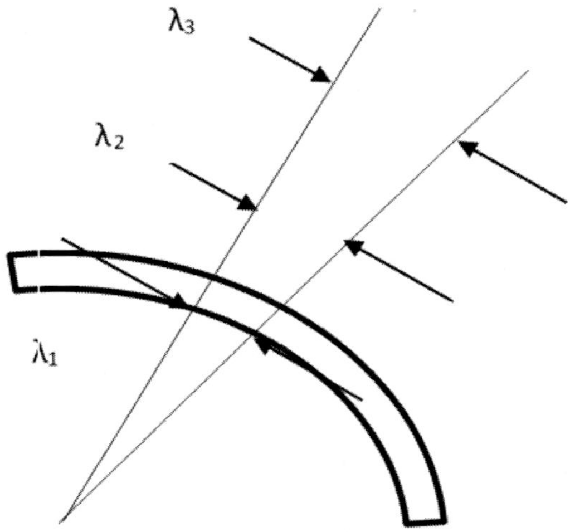

Figure 47. Dependence of the space period produced by the wave propagating in a curve waveguide in the surrounding the waveguide space. As is seen the space period increases with an increase in the distance from the waveguide.

Radiation loss depends on the ratio of the wavelength and the radius of curvature of the waveguide. The radiation losses in an ideal glass ball, in which losses due to molecular light scattering are absent, was calculated [106]. These losses are expressed in the quality factor Q of the resonator, along the surface of which the light wave circulates. The quality factor of the resonator using circulating light varies depending on the number of wavelengths N along the equator on the surface of the ball. At $N = 60$, the quality factor of the resonator $Q= 3\ 10^9$, at $N = 66$ $Q= 3\ 10^{10}$, at $N = 2000$ $Q= 10^{400}$. As is known, the lifetime of oscillations in a resonator is equal to the quality factor multiplied by the oscillation period. Thus, the time of existence of circulating light in an ideal medium without losses is limited due to radiation losses that can vary over a very wide range. Theoretically, this time can vary from several periods to an arbitrarily large one. It is difficult to find other objects whose lifetimes would change within such large limits with a slight change in their parameters.

The behavior of circulating light in the form of Ball Lightning over time is well documented. Ball Lightning does not noticeably change the strength of light and disappears unexpectedly, without a trace, and almost instantly at any part of its trajectory. It is shown that the disappearance of the circulating light is associated with the onset of instability [37]. When the energy of the circulating light falls below a certain threshold, the circulating light becomes unstable. Radiation losses increase, which leads to a decrease in the degree of air compression. In this case, the refractive index of the compressed air decreases, which leads to a further increase in radiation losses, etc. Thus, positive feedback comes into play, which leads to almost instantaneous radiation of the circulating air into the surrounding space. What is left is compressed air, which expands at the speed of sound.

In this case, such instability occurs in a situation that differs only slightly from the stationary state (this state is conditional since the energy of the circulating light is not constant because of radiation). Since Ball Lightning disappears unexpectedly, we can assume that the stored energy at the disappearance is not significantly different from the stored energy at the steady-state. This conclusion is confirmed by the analysis of Ball Lightning stability [43].

The fall of an icicle from the roof is explained in a similar way. The warm water heated by the sun flowing down from the roof gradually melts the base of the icicle and its thickness decreases. When the thickness falls below a certain threshold, the icicle becomes unstable and falls off the roof due to the force of gravity.

A similar situation occurs for a rotating top. As a result of energy losses, the angular velocity of the top gradually decreases, and at a certain speed, instability occurs, as a result of which the top goes into another state.

Note that the steady-state gradually changes and eventually becomes unstable in all given examples.

As for the objects of the micro world in particular the nuclei of elements, the lifetime of nuclei for various elements can change by 10^{50} times. It is not noted that they are subject to any changes during their life. Ultimately there is an unexpected spontaneous decay of the nucleus as a result of natural radioactive decay. The era of nuclear physics began with the discovery of this phenomenon.

In the English Wikipedia, in the Theoretical basis section, this phenomenon is explained in the following scientific way

> "The decay process, like all hindered energy transformations, may be analogized by a snowfield on a mountain. While friction between the ice crystals may be supporting the snow's weight, the system is inherently unstable with regard to a state of lower potential energy. A disturbance would thus facilitate the path to a state of greater entropy; the system will move towards the ground state, producing heat, and the total energy will be distributable over a larger number of quantum states thus resulting in an avalanche. The total energy does not change in this process, but, because of the second law of thermodynamics, avalanches have only been observed in one direction and that is toward the "ground state" — the state with the largest number of ways in which the available energy could be distributed.
>
> Such a collapse (a gamma-ray decay event) requires a specific activation energy. For a snow avalanche, this energy comes as a disturbance from outside the system, although such disturbances can be arbitrarily small. In the case of an excited atomic nucleus decaying by gamma radiation in a spontaneous emission of electromagnetic radiation, the arbitrarily small disturbance comes from quantum vacuum fluctuations."

The above explanation of the cause of the avalanche does not stand up to criticism. The avalanche can be explained on the basis of simple laws of mechanics without any reference to the second law of thermodynamics. Before the avalanche, the snow melts, and the friction force of the snow with the mountain slope decreases. When the friction force becomes less than a certain threshold, the system becomes unstable and the steady-state disappears. The same mechanism takes place when an icicle falls or when the spinning top stops.

If the reference to an avalanche is given as an analogy of radioactive decay, then it should be indicated which process leads to the appearance of instability. What changes in the nucleus cause the nucleus to disintegrate?

If we do not take into account the unfortunate analogy with an avalanche, then from the above description it follows that the nucleus does not change during its life. Quantum vacuum fluctuations are responsible for the decay of the nucleus. It follows from the above explanation that the nucleus is considered to be unchanged during its lifetime, and the vacuum quantum fluctuations bring the nucleus into a state with a different lower energy level, i.e., it decays. The same situation arises with an avalanche if its descent is provoked by a cannon shot. For an avalanche, such a shot is analogous to quantum vacuum fluctuations for a nucleus.

If we assume that, by analogy with circulating light, the stored energy in the nucleus gradually decreases, then we need to conclude that the reason for the decay of the nucleus is the gradual decrease in the energy stored within it. In Ball Lightning, such a decrease in the energy before decay is negligible. It is noted that Ball Lightning suddenly disappears without a visible change in the intensity of the emitted light. It is possible that such a situation takes place in the nucleus, but the measurement of the energy in a single nucleus is not available for experimental verification. Under this assumption, we must admit that nuclear energy is transformed into another form not only during the natural decay of the nucleus but also during the existence of the nucleus. In this case, one can try to explain the wave properties of particles during their passage through a slit by the presence outside the particles of what they radiate during their lifetime.

During the decay of the nucleus, it is known what form of energy arises in this case. Also is known, the type of energy that is gradually radiated by Ball Lightning. In the event that the energy of the nucleus gradually decreases during its existence, the type of energy that the nucleus radiates is unknown. Everything that is unknown to physicists is called dark matter.

We considered Ball Lightning as an example of an object that has the property of wave-particle duality in the macro world. All the properties and behavior of such an object are known since are based on the generally accepted laws of physics and optics. The phenomena responsible for the change in the trajectory of the object when passing through the slit and the reasons for the finite lifetime of the object are shown. The comparison between objects of micro world and macro world shows that these objects have in common that their lifetime can vary by dozens of orders of magnitude. At the same time, the reasons for the decay of the objects of macro world are known and consist in the appearance of instability as a result of a gradual decrease in the stored energy due to inevitable radiation losses. Such losses inevitably arise due to the confinement of the circulating wave to a certain finite volume. In this case, radiation losses arise due to the propagation of the wave along a curved trajectory.

As for the objects of the micro world, it is difficult to determine the cause of radioactive decay from the explanation given in Wikipedia. If we keep in mind the analogy with an avalanche, then the reason for the decay is the appearance of instability. In this case, we must recognize that the cause of radioactive decay is the same as for the circulating light. Changes occur in the nucleus over time are responsible for their instability. However, the cause of the instability is not specified. The presence of wave properties in the objects of the micro word indicates that these objects have some kind of field outside. The interaction of these fields causes the appearance of a force that changes the trajectory of these objects when they are passing through the split. In this case, it can be assumed that motionless objects located at a certain distance interact with each other. If this interaction changes in time periodically, that is, micro-objects are oscillators, then a result of such an interaction can be accumulated in time. In other words, this interaction can be strong. In this case, it can be assumed that the causes of decay for objects of macro world and micro world are identical.

Returning to natural radioactive decay, we must recognize that this phenomenon is a nuclear reaction, the rate of which is extremely low. It is more correct to say that substances for which the rate of radioactive decay was noticeable have long since decayed and disappeared. Substances remained on the Earth with a low rate of radioactive decay. At the same time, substances with a relatively high rate of radioactive decay are created artificially in nuclear reactors. Nuclear reactions connected with the radioactive decay are used to generate energy in cases where such energy is urgently needed, such as in a satellites.

Various elements with different half-lives can be used as a heat source. For example, plutonium 238 has a half-life of 88 years. The power released in this case is 0.54 W/g. A device weighing several kilograms produces a power of several kilowatts and can function autonomously and continuously for several decades. Strontium 90 has a half-life of 10 years. The power released in this case is 0.46 W/g. Polonium 210 has a half-life of 95 days. Polonium weighing 0.5 g is heated under natural conditions to a temperature of 500 degrees Celsius.

Here is a legitimate question. Is it possible to increase the decay rate for elements preserved on Earth? Science answers this question in the negative. The rate of radioactive decay does not depend on either temperature, or pressure, or the type of molecules that include the element in question. However, recently there have been reports that in some cases the rate of decay can be changed. For example, in the US patent US. Pat. no. 5,076,971 Method for enhancing alpha decay in radioactive materials, it has been shown that when metals susceptible to alpha, beta, and gamma radioactivity are placed in an electrical potential of 50 kV to 500 kV, the rate of radioactive decay increases. For example, the rate of radioactive alpha decay of thorium 230 increases 44900 times, and the rate of radioactive beta decay of thallium 204 increases 15 times. Akulov and Mamyrin showed that the half-life of molecular tritium is 11.5 days longer than the half-life of atomic tritium. [122].

How Nuclei Interact with Each Other without Overcoming the Coulomb Barrier?

If we consider nuclei to be corpuscles, then this question is meaningless, and many nuclear physicists will say that only fools can ask this question. This point of view is shared by most specialists in the field of nuclear physics. If we take into account the wave properties of nuclei, or rather the presence of something outside the volume in which the mass of the nucleus is concentrated, then additional analysis is required, presented below.

If we keep in mind wave particle duality and the similarity between the disappearance of circulating light and the radioactive decay of nuclei, then we can assume that the wave properties of nuclei are manifested due to the waves circulating in the nucleus. In this case, the nuclear can be imagined as a conservative oscillatory system with natural frequencies.

Unlike corpuscles, a noticeable interaction between which is characterized by a short interaction time, but a significant influence of corpuscles on each other, a noticeable interaction between oscillatory systems can be achieved in a completely different way. To do this, it is sufficient to provide long-term interaction with a relatively weak degree of interaction. In this case, there is no need to bring the nuclei closer together at a small distance. However, it is necessary that the position of two identical interacting nuclei remains unchanged for a relatively long time.

We have shown that natural radioactive decay can be explained on the basis of the wave theory under the assumption that the nucleus is an oscillator in which the stored energy is gradually reduced due to the radiation loss. It was assumed that the nucleus becomes unstable and decays when the stored energy becomes less than a certain threshold. We gave an analog of optics in a form of circulating light where self-confined light has the same properties. Now we are going to consider physical phenomena that can increase the rate of radioactive decay.

There is a fundamental difference between the interaction of corpuscles and the interaction of oscillators. In order to obtain a strong interaction between the corpuscles, in which nuclear transformations are possible, it is necessary that the corpuscles have a significant kinetic energy, that is, they must move relatively to each other with velocities close to the velocities of light, as occurs in micro particle accelerators.

The nature of the interaction between the oscillators is completely different. An interaction can occur even between motionless oscillators. Even with a weak coupling between adjacent oscillators may have a significant result during long-term interaction. The long-term interaction is not a problem for nuclei that exist for a long time.

The self-confined light where waves are circulating are oscillators. The frequencies of the waves are the own frequencies of oscillators. The oscillators in a form of nuclei are excellent devices for exchange of energies between them, since the lifetime of the accumulated energy is much greater than the time during which the exchange takes place. Thus, if one nucleus is placed in a place where an evanescent field of the other nucleus is different from zero, then there is an interaction between such nuclei. The strength of this interaction depends on how much the evanescent field of one nucleus penetrates into the evanescent field of the other nucleus. Since the amplitude of the wave decreases with increasing distance from the nucleus, the interaction degree is maximal between adjacent nearby nuclei the distance between them is as small as possible. In addition, the amplitude of the wave outside the nucleus is the greater, the shorter the half-life of the nuclei. In this case, nuclei with a small half-life (including radiation waste) should interact with each other better.

The theory of coupled oscillators is well developed and is used in many completely different applications. We give its main results, which are used in the subsequent consideration. This consideration is greatly facilitated if identical lossless oscillators are considered. It can be assumed that the natural frequencies of identical nuclei are the same, and the oscillations do not damp out, since the nuclei exist much longer than the interaction time. Further, we will consider the interaction only between identical nuclei, since otherwise the nuclei have different natural frequencies and there is practically no interaction between such nuclei.

Two identical weakly coupled oscillators are characterized by the following two parameters: the natural frequencies ω and $\Delta\omega \ll \omega$. Here ω is the natural frequency of an isolated oscillator and parameter $\Delta\omega$ characterizes a degree of coupling. A system of two identical coupling resonators has two natural solutions: symmetrical and anti-symmetrical. Amplitudes and phases of oscillations are identical for the symmetrical one. Amplitudes of oscillations are identical for the anti-symmetrical one but phases of oscillation are differed by π.

With a symmetric own solution, the oscillations have a frequency ω-$\Delta\omega$. For an antisymmetric own solution, oscillations have a frequency ω+$\Delta\omega$. Since the stored energy in an oscillator varies proportionally to its natural frequency with slow (adiabatic) changing the oscillator parameters, the symmetric solutions has a minimum energy. The system of coupled oscillators tends to minimize the stored energy. The minimum decreases with increasing the degree of coupling $\Delta\omega$. Consequently, the oscillators tend to approach each other, that is, they are attracted to each other if the oscillations in them are in phase and repel each other if the phase of oscillations differs by π.

As is known, the lifetime of a neutron in a free state is about 15 minutes. In addition, it is known that the neutron has wave properties. Therefore, in accordance with the above arguments, the neutron is an oscillatory system. In the nucleus, neutrons are densely packed and are in a stationary state with a minimum of the energy, that is, the oscillations in them are symmetrical and a phase difference is zero/ The natural frequency of coupled neutrons is equal ω-$\Delta\omega$, where $\Delta\omega$ is a change in the natural frequency as a result of the interaction of all neutrons. Since the energy of an oscillatory system is proportional to the frequency, the mass defect of the nucleus is equal to the mass of all isolated neutrons having their own frequency ω, multiplied by the value of $\Delta\omega/\omega$. Perhaps this phenomenon is the reason for the appearance of forces that bring neutrons closer together, which in the nuclear physics were classified as an independent type of force and called the forces of strong interaction.

Any combination of symmetric and antisymmetric solutions is also a solution. As an example, a solution is possible in which the symmetric and antisymmetric solutions are summed. In this case, the amplitudes of these solutions are the same, and the phases are the same for the first oscillator. In this case, the amplitude in the first oscillator doubles, and the amplitude in the second oscillator becomes zero. Since oscillations with different frequencies are summed up, beats occur in the first oscillator. A similar situation takes place in the second oscillator. As a result, the maximum amplitude in the first oscillator gradually decreases, while in the second one gradually increases. In this case, the energy from the first oscillator is gradually transferred to the second one Then the energy from the second oscillator is pumped into the first, and so on. Thus, there are beats with a frequency $\Delta\omega$. The stronger the coupling between the oscillators, the higher the beat frequency. As an example, a change in time of the energy stored in the oscillator s is shown in Figure 48.

What happens when oscillators with the same energies approach each other? This situation corresponds to point of intersection curve 1 and 2 at t=0 in Figure 48. As can be seen from the figure, the energy from one oscillator can be transferred to another. The rate of the transfer is maximal if the difference phase between oscillations is equal to $\pi/2$ or - $\pi/2$. If the difference of phases is equal 0 or π, there is no transfer. If the phase difference has other values, then a partial energy transfer occurs. The closer the phase difference is to $\pi/2$ or - $\pi/2$, the more energy is transferred between the oscillators. In this case, energy is transferred to the oscillator, the phase of which is delayed.

As can be seen from the Figure 48, energy can be completely transferred from one oscillator to another if oscillators are identical and the initial phase difference between them is equal to $\pi/2$ or - $\pi/2$.

With regard to oscillators in the form of nuclei, this is impossible. We believe that as soon as the energy in the nucleus drops below a certain threshold, the nucleus will move into a new state. What this state is we do not know. Perhaps this is the state that occurs during radioactive decay.

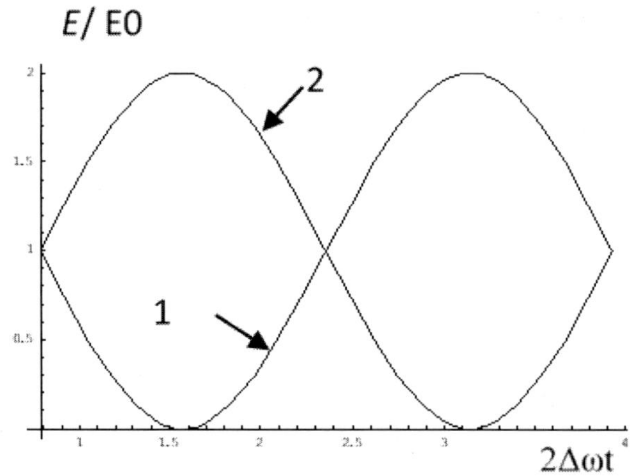

Figure 48. A theoretical relative change of energies stored in resonator 1 (curve 1) and resonator 2 (curve 2) in time t.

However, another situation is also possible, in which the transmutation of identical nuclei occurs. Nuclei can exchange nucleons with minimal loss of total energy. In chemistry, an analogue of such a reaction is the exchange of different atoms between identical molecules. For example, two identical molecules of hydrochloric acid 2HCl exchange in such a way that one molecule gives the other one a chlorine atom, and receives a hydrogen atom in return. As a result, one molecule has two chlorine atoms Cl_2, and the other has two hydrogen atoms H_2.

There is a legitimate question. Why, in this case, the nuclei do not interact in solids and liquids, although the atoms are densely packed and their mutual arrangement does not change in time? The answer to this question is this. The system is in a stationary state with a minimum of energy. The minimum occurs when the oscillations in all nuclei are in phase. If this condition is not met at the moment of fluid formation, then the beat shown in Figure 48 takes place. These beats are accompanied by the emergence of forces that periodically tend to bring the nuclei together and push them away with the beat frequency. As a result, energy is consumed during nuclear vibrations. The total energy of the interacting nuclei decreases until it reaches a minimum. This occurs when the vibrations of the nuclei are in phase.

As confirmation, you can watch a video https://youtu.be/Bt_rUnpT7Jc that shows the synchronization of oscillatory systems in the form of metronomes. Many metronomes, the phases of which are arbitrary at the initial moment of time, begin to oscillate in the same phases over time. On closer examination, you can see that the metronomes are located on an elastic base, through which the vibrations of one metronome are transmitted to the neighboring ones, that is, the metronomes are coupled through the elastic base. In a similar way, oscillations are synchronized in identical nuclei that are motionless relative to each other, the connection between which is carried out through evanescent fields located outside the nuclei.

Thus, the transfer of energy from one nucleus to another can only be expected in a non-stationary transition process, when gas nuclei with random phases become densely packed during condensation of vapor into a liquid. However, this process is quite lengthy and cannot be repeated with a high frequency.

Much more attractive is the process in which the gas is compacted to the density of a liquid with a strong compression of the gas. This process can be carried out by compressing the gas into thin films using the circulating light. At the same time, the time for creating such films is minimal, since the movement of molecules over a minimum distance is required. Since the speed of light is six orders of magnitude greater than the speed of sound, the speed of transient processes in the circulating light increases accordingly.

Thus, from the above considerations, it follows that circulating light is the most suitable candidate for the implementation of nuclear transformations. We presented this conclusion as a logical result based on existing knowledge. In fact, the above arguments can be considered as a justification for the well-known phenomenon in which nuclear transformations and circulating light are quite often mentioned together.

Physical Conditions Favorable for Nuclear Transformation

As follows from the presented theory of coupled oscillators, an increase in the energy transfer rate can be achieved with the closest possible location of oscillators. Considering the nuclei as oscillators, we can conclude that the they should be located as close to each other as possible. Keeping in mind that the distance between the nuclei is much greater than their size and the coupling between the nuclei is rather small, we must conclude that the interaction between the nuclei must be sufficiently long. We do not know the eigen frequencies of the nuclei, therefore we cannot estimate the required time of interaction. We can only point out that this time must be many orders of magnitude greater than the time of interaction of nuclei as corpuscles.

These conditions are satisfied in solids and liquids, however, no nuclear transformations are observed, since in this case the nuclei are in a stationary state with a minimum of total energy. For the transfer of energy from one nucleus to another, it is necessary that the phases of oscillations in neighboring nuclei differ by $\pi/2$. This can only be done in a non-stationary state during transition processes to the steady state.

Taking into account the conditions under which nuclear reactions are observed, we can assume that such a situation occurs when the gas is compressed sufficiently quickly to such a degree that its molecules become densely packed for some time. In this case, we assume that the nuclear phases of oscillation in the gas are randomly distributed due to numerous short-term collisions, during which a small random phase shift occurs in the colliding nuclei.

As a result of such compression, there is a non-zero probability that the phase shift between adjacent nuclei may turn out to be equal to $\pi/2$ in some places, which will lead to a nuclear reaction. After that, the system passes into a stationary state, and the process must be repeated to obtain the next nuclear reactions. In this case, the number of favorable situations is proportional to the frequency of repetitions.

If it is intended to extract energy, the energy expended for periodic compression and expansion of the gas should be less than the energy released during nuclear transformation. If it is intended to destroy radioactive waste, this condition may not be met.

To obtain the maximum output thermal power, it is necessary to increase the frequency of periodic compressions and extensions. Consequently, the compression and expansion times should be as small as possible to increase the output energy, on one side, and great enough to provide required time of interaction between adjacent nuclei on the other side.

Perhaps this approach is used by some bacteria to extract nuclear energy for their vitality. Unlike most bacteria, which extract energy from chemical reactions, as evidenced by a change in the chemical composition of the waste compared to the food they process, bacteria are known whose waste indicates a change in the nuclei of the food they process [121]. There is no high pressure, no high temperature, no intense light, no high electric and magnetic fields. Processing takes place under normal conditions.

It can be assumed that the bacterium simply brings two identical atoms together at the minimum possible distance and waits for a nuclear reaction to occur with the release of energy. Perhaps the bacterium senses which atoms to bring together. The most promising are those atoms that do not attract or repel each other. In this case, phases of oscillations differ by $\pi/2$ and the rate of energy transfer from one nucleus to another is maximum.

How Realize Nuclear Transformation

Consider now the means that provide a performance of the conditions listed above. To solve this problem, the most suitable are the means in which the light participates. The time constant of transient processes τ involving the light is determined by the light speed, whereas the time constant with the participation of mechanical means is determined by the sound speed at best. As is known, in this case, $\tau_{sound} = 100\ 000\tau_{light}$. When the light interacts with an optical medium, the phenomenon called "optical electrostriction pressure" is responsible for the compression of the optical medium.

The optical electrostriction pressure is poorly studied and it is assumed that it can be neglected when considering optical phenomena at commonly used light intensities. This is true for the light propagating linearly. However, for the circulating light, its intensity increases by 10^9 -10^{12} times, depending on conditions of production of objects with the circulating light. For example, a high electrostriction pressure in a shell of the ball light provides a strong compression of air in the shell, at which molecular light scattering decreases by millions times because fluctuations in the air density disappear and molecules are tightly packed and cannot move relative to each other.

We can conclude that the objects with circulating light are ideal candidates for fulfilling the above conditions. At the same time, it is desirable to have a technology when many such objects simultaneously appear with the life-time sufficient to transfer energy between neighboring nuclei.

It turns out that such technologies have been used for a long time by many experimenters who noted the appearance of excess energy. Various theories are used to explain this phenomenon. However, no theory indicates ways to increase its effectiveness. We do not know the types of waves circulating in the nuclei, their own frequencies, the dependence of the degree of coupling on distance between the nuclei. However, only from qualitative considerations, assuming that the nuclei are non-linear oscillators that can go into another state when the oscillation energy decreases below a certain threshold, we were able to determine the conditions that ensure the maximum output of excess energy.

There are quite a few reports where low-energy nuclear transformations are observed when objects with circulating light arises. Different authors call these objects differently.

Avramenko calls them abnormal objects [116]
Schoulders calls them exotic vacuum objects (EVO) [117]
Mesyats calls them ectons [118]
Lewis calls them miniature Ball Lightning [119]
Urutskoev calls them spherical formations or strange radiation [120].

We will call them bubbles of light or ball light.

It is not a place where all sources should be listed. The method of obtaining such objects is almost the same for all the authors listed. This is a powerful electrical discharge, where an erosion of matter from the electrodes is accompanied by an intense radiation of light. The substance passes into a gaseous state. However, there is no generally accepted theoretical justification for the observed phenomena. This is evidenced by the definitions given to the received objects - anomalous, exotic, strange. Lewis calls them miniature Ball Lightning.

We believe that Lewis is right. Indeed, all these objects are special cases of a wide class of objects, to which also natural Ball Lightnings belong. We have shown that Ball Lightning is the self-confined light appeared because of the instability of intense light in gas [13]. We can conclude about an existence of a large pressure in the shell from the fact that the lifetime of a light bubble is a fraction of a second, while the lifetime of light in a common terrestrial atmosphere is several milliseconds because of the molecular light scattering due to inhomogeneity of the reflective index arising from air density fluctuations. At a sufficiently strong pressure, when the gas molecules are tightly packed, the density fluctuations disappear, which leads to an increase in the lifetime of the light. There are densely packed molecules in the shell of the circulating light that are required to accelerate nuclear transformation.

Also another way to quickly generate a very high pressure is known. These phenomena occur in the cavitation bubbles of the liquid. Cavitation (from Latin cavitas - emptiness) - the appearance of a cavity filled with vapors gases or a mixture of them in a liquid. Cavitation bubbles appear when the pressure of a liquid is reduced to such an extent that the liquid begins to boil at this pressure. If the pressure decreases due to an increase in its velocity, then this phenomenon is called hydrodynamic cavitation. The physical properties of cavitation bubbles were considered earlier.

Usually water is used as the liquid. The resulting cavity in the water takes the form of a hollow sphere due to the surface tension forces. This sphere is gradually filled with liquid vapors and gases dissolved in the liquid. Unlike liquid vapors, dissolved gases have a high saturation vapor pressure. Therefore, they do not condense when the pressure in the sphere increases.

A diameter of the cavitation bubble is usually several tens microns. Since the curvature of the bubble surface is very large, there is a large surface pressure that tends to reduce the bubble diameter. The pressure can be over 10,000 atmospheres. In this case, the bubble diameter rapidly decreases, and the gas pressure inside the bubble increases as it is compressed. As the pressure increases, part of the water vapor turns into a liquid water with the release of heat. In this case, an adiabatic process takes place. According to the Poisson equation, the gas temperature increases with decreasing volume. As shown experimentally, the temperature can reach 6000 C^0.

Currently, there are quite a lot of mass-produced devices that use the phenomenon of cavitation to obtain additional energy due to additional heating of water during its cavitation. Only in Russia hundreds of devices with power from units to hundreds of kilowatts are operated. Currently, heat generators are already being mass-produced.

Thus, the excess energy generated during cavitation is an objective reality, although there is currently no generally accepted scientific explanation for this phenomenon. The above detailed data and analysis of the properties of cavitation bubbles suggest that the excess energy is associated with accelerated nuclear transformation.

The following observation confirms our conclusions. It is noticed that the effect of water heating by cavitation bubbles gradually weakens if the heated water is then used again in a closed cycle. We explain this phenomenon by the fact that in this case the number of nucleus in water that release heat during transformation gradually decreases, since the transformation of nuclei is accompanied by their change.

Abnormal Great Life Time of Circulating Light

Another unique property of the circulating light is the extremely long lifetime in an optical medium. Let us show that this property of the circulating light is a consequence of the extremely high pressure of the air in the shell in which it circulates. This is an additional evidence of the results obtained in determining the necessary conditions for the stability of circulating light [46].

As is known, a duration of an existence of natural Ball Lightnings lasts from several seconds till several minutes. This is significantly greater than the life time of conventional white light in the terrestrial atmosphere. Indeed, the sun light decreases by 1.3% at a distance of 1 km in the terrestrial atmosphere. Bearing in mind that the speed of light is 300 000 km/s, we find that the light decreases 1.5 times per 100 μs. This is 4-5 orders of magnitude smaller than the life time of Ball Lightning.

It is known that the phenomenon of molecular light scattering is responsible for decreasing intensity of the light in the atmosphere

But the scattering of light in the shell of ball light differs in its character from the scattering of light in the Earth's atmosphere [7]. The light continues to remain in the shell after scattering and propagates in different directions, in contrary to the scattering of light in the Earth's atmosphere, where the scattered light is irretrievably lost. The directional pattern of the scattered light is shown in Figure 49.

As follows from Figure 49, the light is scattered to the greatest extent in the directions coinciding with the incident light, or in the direction opposite to the incident light. Light is weakly scattered in directions perpendicular to the incident light. Rough estimates show that this effect can increase the lifetime of light in a ball light by about an order of magnitude.

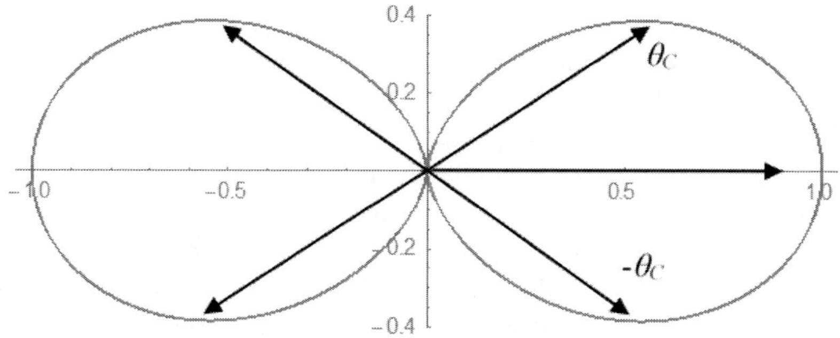

Figure 49. Dependence of the scattering light on the angle θ between a direction of incident light and direction of the scattering light. As is seen, the scattering is maximal along the incident direction.

Dependence of the intensity of the scattering light on the angle of scattering θ is presented in Figure 50. A vertical line separates angles of scattering where the scattering light continues to propagate in the lightguide (left part) and the angles of scattering where the scattering light can be radiated in surrounding space. We need to note that in accordance with the Fresnel laws only a part of this light leaves the lightguide. Another part returns to the lightguide due to the reflection. For us, it does not make sense to consider in more detail, since it is necessary first to determine the parameters of the lightguide from compressed gas.

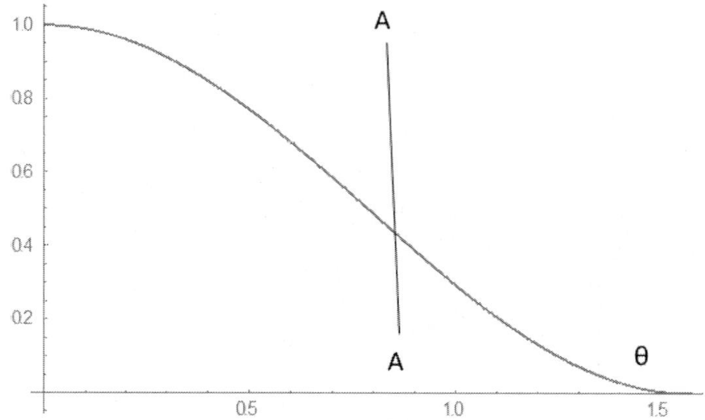

Figure 50. Relative ratio between scattering waves that continue to propagate in the waveguide due to the total inner reflections (left area from line AA) and scattering waves that penetrate partially in free space (right area from line AA).

We can conclude that a difference between the molecular light scatterings in a homogeneous air atmosphere and the lightguide formed by the compressed air can be significant. A gas pressure about 21-25 GPa was mentioned early [7]. From presented consideration this magnitude can be decreased till 2.1- 2.5 GPa. This magnitude agrees with that obtained at consideration of conditions of stability [46].

However, we must also explain the other orders of magnitude by which the lifetime of light in the ball light exceeds the lifetime of light in the Earth's atmosphere. We have shown in the very first work on the optical nature of Ball Lightning [7] that a decrease of the scattering light is possible at significant decrease a fluctuation of the air density that are responsible for the scattering of light in atmosphere.

At very great compression about σ=1000, the compressibility of the air decreases by a factor above 2 10^4 times. This small compressibility prevents arising the air density fluctuation because at any attempt to occasionally form an increase in the air density, the force that prevent compression come into play due to extremely small compressibility of the air at such a small air compression. Besides, the circulating light of great energy density also prevents arising fluctuations.

Thus, the circulating light is a suitable candidate for creating conditions for nuclear transformations, since the physical conditions produced by the circulating light correspond to the above conditions necessary for interaction between identical nuclei with each other.

Possible Nuclear Transformations during Radioactive Decay

As was noted earlier, there is an opinion that the act of natural radiative decay takes place when the energy of a certain nucleus becomes less than a certain threshold. In this case, the nucleus becomes unstable and ceases to exist. The interaction of two identical nuclei, considered as coupled oscillators, causes a much faster decrease in the energy stored in one of nuclei.

In the same time, a completely different interaction is possible, in which the second nucleus also changes. In conventional chemistry, the breakdown of a molecule occurs as a result of a decay reaction. However, in chemistry, substitution reactions are more common, in which molecules are exchanged by atoms. For example, hydrochloric acid and caustic sodium exchange atoms of chlorine, oxygen and hydrogen. The result is two completely different molecules - molecules of water and sodium chloride. It is possible that similar substitution reactions are also possible in nuclear reactions where nuclei are changed by neutrons and protons rather than by atoms in chemical reactions. At present, such nuclear reactions are commonly called nuclear transmutation.

Molecules are very diverse and can consist of a different number of various atoms, of which there are currently about 100. Nucleons in nuclear reactions serve as an analogue of atoms in chemical reactions. There are only two types of nucleons known: protons and neutrons. The number of different nuclei is limited and, together with isotopes, does not exceed several thousand. It is possible that nuclear reactions between different nuclei are possible. We do not know the mechanism of these reactions and therefore do not consider. We have shown that reactions between two identical nuclei are possible. Therefore, we consider only them.

An analogue of such a reaction in chemistry is the reaction between two molecules of hydrochloric acid HCl. One molecule donates a chlorine atom to another molecule and takes a hydrogen atom. The result is two different molecules, one consisting of two chlorine atoms Cl_2 and the other of two hydrogen atoms H_2.

We assume that the smaller the excess energy, the smaller the potential barrier between the states before and after the transformation and the greater its probability. The transformation may occur if the excess energy is positive, that is, the difference between the total energies of the nuclei before and after transformation is positive.

As an example, we present a table of possible nuclear substitution reactions. In this case, reactions were chosen in which excess energy is released, and the reaction products are stable isotopes. There are about 550 such reactions in total. The table lists some of the reactions in which the excess energy is less than 1 MeV. The table consists of 14 columns.

Columns 1 indicates a name of the element whose two identical nuclei enter into a substitution reaction.

Columns 2 indicates the total number of nucleons (a sum of protons and neutrons) in the nucleus.

Columns 3 indicates the serial number of the element (a number of protons) in the nucleus.

Columns 4 indicates the defect mass of the element taken from handbook.

Columns 5indicates a name of the first element obtained as a result of nuclear transformation.

Columns 6 indicates the total number of nucleons (a sum of protons and neutrons) in the first element.

Columns 7 indicates the serial number of the element (a number of protons) in the first element.

Columns 8 indicates the defect mass of the first element taken from handbook.

Columns 9 indicates a name of the second element obtained as a result of nuclear transformation.

Columns 10 indicates the total number of nucleons (a sum of protons and neutrons) in the second element.

Columns 11 indicates the serial number of the element (a number of protons) in the second element.

Columns 12 indicates the defect mass of the second element taken from handbook.

Columns 13 indicates the excess energy obtained as a result of the nuclear transformation.

Columns 14 indicates a unit of the presented energy.

Strings in the table are sorted in ascending order of excess energy (column 13).

From this table, you can see that for each string describing a particular reaction, the number of protons before and after the reactions is stored. The same is true for the number of nucleons. Then a number of neutrons before and after the reactions is stored also. As a result of the reaction, only excess energy is released.

Note that the mass defects of the initial nuclei and the mass defects of the nuclei formed as a result of the reaction significantly exceed the energy obtained as a result of the reaction.

In addition, in all the reactions presented, the mass defect of one of the emerging nuclei becomes smaller than the initial one. Then the mass defect of another emerging nucleus becomes greater than the initial one. This means that the neutrons and protons are more densely packed in this nucleus than in the original one. Therefore, in such reactions, there is no need to insert into the nucleus the energy exceeding the mass defect in order to overcome the forces that keep a nucleus from decay.

Table 2. Possible nuclear reaction at acceleraticn of radioactive decay

1	2	3	4	5	6	7	8	9	10	11	12	13	14
Copper	34	29	65578	Germanium	38	32	70561	Iron	30	26	60604	9	keV
Technetium	54	43	87224	Cadmium	60	48	89951	Strontium	48	38	84512	15	keV
Mercury	121	80	27672	Thallium	124	81	25769	Gold	118	79	29591	16	keV
Palladium	59	46	88422	Cadmium	63	48	89254	Ruthenium	55	44	87619	29	keV
Ruthenium	57	44	87951	Silver	62	47	88722	Niobium	52	41	87209	29	keV
Magnesium	14	12	16212	Silicon	16	14	24431	Neon	12	10	8026	33	keV
Cesium	78	55	88089	Lanthanum	82	57	87231	Iodine	74	53	88981	34	keV
Manganese	30	25	57710	Iron	31	26	60179	Chromium	29	24	55283	42	keV
Thallium	122	81	25988	Lead	125	82	22463	Mercury	119	80	29557	44	keV
Gadolinium	91	64	72071	Terbium	94	65	69536	Europium	88	63	74651	45	keV
Zirconium	52	40	84822	Ruthenium	58	44	89100	Krypton	46	36	80591	47	keV
Thulium	100	69	61269	Ytterbium	101	70	59302	Erbium	99	68	63286	50	keV
Gadolinium	90	64	73704	Dysprosium	92	66	70411	Samarium	88	62	77049	52	keV
Osmium	113	76	38978	Gold	118	79	29591	Tantalum	108	73	48425	60	keV
Tin	67	50	90398	Tellurium	72	52	90518	Cadmium	62	48	90349	71	keV
Chromium	29	24	55283	Iron	32	26	62151	Titanium	26	22	48487	72	keV
Palladium	62	46	89523	Cadmium	66	48	90019	Ruthenium	58	44	89100	73	keV
Silver	62	47	88722	Tellurium	68	52	89404	Molybdenum	56	42	88115	75	keV
Titanium	27	22	48558	Iron	32	26	62151	Argon	22	18	35040	75	keV
Technetium	56	43	87326	Palladium	60	46	89913	Zirconium	52	40	84822	83	keV
Tin	67	50	90398	Tellurium	70	52	90304	Cadmium	64	48	90577	85	keV
Holmium	98	67	64896	Erbium	102	68	60104	Dysprosium	94	66	69774	86	keV
Molybdenum	53	42	87712	Ruthenium	55	44	87619	Zirconium	51	40	87892	87	keV
Mercury	119	80	29557	Mercury	122	80	27356	Mercury	116	80	31846	88	keV
Titanium	25	22	44931	Iron	32	26	62151	Argon	18	18	27799	88	keV
Dysprosium	97	66	66382	Erbium	98	68	64921	Gadolinium	96	64	67943	100	keV
Germanium	40	32	72582	Selenium	42	34	75259	Zinc	38	30	70006	101	keV

Table 2. (Continued)

1	2	3	4	5	6	7	8	9	10	11	12	13	14
Krypton	44	36	77897	Strontium	46	38	80641	Selenium	42	34	75259	106	keV
Lutetium	104	71	55159	Hafnium	105	72	52879	Ytterbium	103	70	57546	107	keV
Hafnium	107	72	50462	Tungsten	109	74	46347	Ytterbium	105	70	54691	114	keV
Ruthenium	58	44	89100	Palladium	62	46	89523	Molybdenum	54	42	88794	117	keV
Selenium	46	34	77761	Krypton	48	36	82431	Germanium	44	32	73213	122	keV
Selenium	42	34	75259	Strontium	46	38	80641	Zinc	38	30	70006	129	keV
Neodymium	86	60	80923	Gadolinium	90	64	73704	Barium	82	56	88273	131	keV
Molybdenum	56	42	88115	Ruthenium	58	44	89100	Zirconium	54	40	87263	133	keV
Platinum	116	78	34765	Mercury	118	80	30964	Osmium	114	76	38699	133	keV
Germanium	38	32	70561	Selenium	42	34	75259	Zinc	34	30	66001	138	keV
Erbium	99	68	63286	Ytterbium	104	70	56940	Dysprosium	94	66	69774	142	keV
Arsenic	42	33	73033	Bromine	44	35	76070	Gallium	40	31	70141	145	keV
Holmium	98	67	64896	Ytterbium	102	70	59250	Gadolinium	94	64	70691	149	keV
Erbium	99	68	63286	Ytterbium	100	70	60759	Dysprosium	98	66	65967	154	keV
Zinc	38	30	70006	Germanium	42	32	73422	Nickel	34	28	66745	155	keV
Gadolinium	92	64	72536	Dysprosium	96	66	68181	Samarium	88	62	77049	158	keV
Titanium	24	22	44122	Iron	30	26	60604	Argon	18	18	27799	159	keV
Tellurium	73	52	89019	Iodine	74	53	88981	Antimony	72	51	89217	160	keV
Molybdenum	52	42	88412	Ruthenium	54	44	88226	Zirconium	50	40	88764	166	keV
Tellurium	68	52	89404	Xenon	70	54	87450	Tin	66	50	91526	168	keV
Iodine	74	53	88981	Barium	81	56	87733	Tin	67	50	90398	169	keV
Palladium	58	46	89400	Tin	64	50	90561	Molybdenum	52	42	88412	173	keV
Palladium	56	46	87925	Cadmium	60	48	89951	Ruthenium	52	44	86075	176	keV
Gallium	38	31	69321	Selenium	43	34	74601	Nickel	33	28	64219	178	keV
Tin	65	50	90035	Tellurium	70	52	90304	Cadmium	60	48	89951	185	keV
Osmium	112	76	41125	Mercury	119	80	29557	Hafnium	105	72	52879	186	keV
Dysprosium	95	66	68056	Ytterbium	102	70	59250	Samarium	88	62	77049	187	keV

1	2	3	4	5	6	7	8	9	10	11	12	13	14
Calcium	23	20	38405	Calcium	24	20	41466	Calcium	22	20	35543	199	keV
Tellurium	70	52	90304	Xenon	72	54	89162	Tin	68	50	91653	207	keV
Copper	36	29	67261	Germanium	40	32	72582	Iron	32	26	62151	211	keV
Lutetium	104	71	55159	Hafnium	108	72	49779	Ytterbium	100	70	60759	220	keV
Oxygen	9	8	809	Magnesium	13	12	13190	Beryllium	5	4	-11348	224	keV
Cadmium	62	48	90349	Tin	66	50	91526	Palladium	58	46	89400	228	keV
Antimony	70	51	89588	Tellurium	73	52	89019	Tin	67	50	90398	241	keV
Silver	62	47	88722	Tin	69	50	90066	Ruthenium	55	44	87619	241	keV
Tin	65	50	90035	Tin	68	50	91653	Tin	62	50	88658	241	keV
Cesium	78	55	88089	Barium	81	56	87733	Xenon	75	54	88697	252	keV
Ruthenium	60	44	88099	Palladium	64	46	88335	Molybdenum	56	42	88115	252	keV
Platinum	116	78	34765	Thallium	122	81	25988	Rhenium	110	75	43802	260	keV
Dysprosium	95	66	68056	Dysprosium	98	66	65967	Dysprosium	92	66	70411	266	keV
Molybdenum	54	42	88794	Ruthenium	58	44	89100	Zirconium	50	40	88764	276	keV
Xenon	75	54	88697	Barium	82	56	88273	Tellurium	68	52	89404	283	keV
Hafnium	105	72	52879	Osmium	112	76	41125	Erbium	98	68	64921	288	keV
Technetium	55	43	86434	Molybdenum	55	42	85544	Ruthenium	55	44	87619	295	keV
Technetium	55	43	86434	Ruthenium	55	44	87619	Molybdenum	55	42	85544	295	keV
Hafnium	106	72	52434	Tungsten	108	74	48228	Ytterbium	104	70	56940	300	keV
Hafnium	107	72	50462	Osmium	112	76	41125	Erbium	102	68	60104	305	keV
Cadmium	63	48	89254	Tin	67	50	90398	Palladium	59	46	88422	312	keV
Mercury	121	80	27672	Mercury	124	80	24703	Mercury	118	80	30964	323	keV
Rhodium	58	45	88024	Palladium	59	46	88422	Ruthenium	57	44	87951	325	keV
Strontium	48	38	84512	Zirconium	50	40	88764	Krypton	46	36	80591	331	keV
Magnesium	13	12	13190	Aluminum	14	13	17194	Sodium	12	11	9529	343	keV
Thulium	100	69	61269	Ytterbium	104	70	56940	Erbium	96	68	65941	343	keV
Osmium	113	76	38978	Iridium	116	77	34519	Rhenium	110	75	43802	365	keV
Palladium	58	46	89400	Cadmium	60	48	89951	Ruthenium	56	44	89221	372	keV
Cesium	78	55	88089	Xenon	80	54	88125	Barium	76	56	88453	400	keV

Table 2. (Continued)

1	2	3	4	5	6	7	8	9	10	11	12	13	14
Platinum	114	78	36283	Mercury	116	80	31846	Osmium	112	76	41125	405	keV
Tin	69	50	90066	Tellurium	72	52	90518	Cadmium	66	48	90019	405	keV
Molybdenum	53	42	87712	Palladium	56	46	87925	Strontium	50	38	87910	411	keV
Cadmium	64	48	90577	Tin	68	50	91653	Palladium	60	46	89913	412	keV
Magnesium	13	12	13190	Sulfur	16	16	26015	Oxygen	10	8	783	418	keV
Samarium	90	62	74761	Gadolinium	92	64	72536	Neodymium	88	60	77407	421	keV
Ruthenium	57	44	87951	Ruthenium	60	44	88099	Ruthenium	54	44	88226	423	keV
Samarium	92	62	72454	Gadolinium	96	64	67943	Neodymium	88	60	77407	442	keV
Tellurium	73	52	89019	Xenon	77	54	88421	Tin	69	50	90066	449	keV
Palladium	60	46	89913	Tin	66	50	91526	Molybdenum	54	42	88794	494	keV
Tellurium	72	52	90518	Xenon	76	54	89881	Tin	68	50	91653	498	keV
Ytterbium	105	70	54691	Hafnium	108	72	49779	Erbium	102	68	60104	501	keV
Dysprosium	95	66	68056	Erbium	96	68	65941	Gadolinium	94	64	70691	520	keV
Platinum	118	78	32652	Mercury	124	80	24703	Osmium	112	76	41125	524	keV
Samarium	88	62	77049	Gadolinium	90	64	73704	Neodymium	86	60	80923	529	keV
Tungsten	109	74	46347	Platinum	114	78	36283	Ytterbium	104	70	56940	529	keV
Chlorine	20	17	31761	Argon	22	18	35040	Sulfur	18	16	29013	531	keV
Rhodium	58	45	88024	Indium	64	49	89372	Niobium	52	41	87209	533	keV
Rubidium	48	37	82158	Strontium	49	38	84868	Krypton	47	36	79984	536	keV
Zirconium	51	40	87892	Molybdenum	52	42	88412	Strontium	50	38	87910	538	keV
Indium	64	49	89372	Tin	65	50	90035	Cadmium	63	48	89254	545	keV
Indium	64	49	89372	Tellurium	68	52	89404	Palladium	60	46	89913	573	keV
Gadolinium	93	64	70825	Dysprosium	94	66	69774	Samarium	92	62	72454	578	keV
Germanium	41	32	71293	Arsenic	42	33	73033	Gallium	40	31	70141	588	keV
Technetium	54	43	87224	Cadmium	58	48	87131	Strontium	50	38	87910	593	keV
Calcium	24	20	41466	Titanium	26	22	48487	Argon	22	18	35040	595	keV
Terbium	94	65	69536	Dysprosium	98	66	65967	Gadolinium	90	64	73704	599	keV

1	2	3	4	5	6	7	8	9	10	11	12	13	14
Platinum	117	78	32802	Platinum	120	78	29921	Platinum	114	78	36283	600	keV
Palladium	59	46	88422	Palladium	62	46	89523	Palladium	56	46	87925	604	keV
Terbium	94	65	69536	Erbium	98	68	64921	Samarium	90	62	74761	610	keV
Iridium	114	77	36698	Platinum	117	78	32802	Osmium	111	76	41208	614	keV
Lead	125	82	22463	Lead	126	82	21758	Lead	124	82	23795	627	keV
Selenium	42	34	75259	Krypton	46	36	80591	Germanium	38	32	70561	634	keV
Gallium	38	31	69321	Selenium	40	34	72212	Nickel	36	28	67097	667	keV
Oxygen	9	8	809	Aluminum	14	13	17194	Lithium	4	3	-14908	668	keV
Ruthenium	56	44	89221	Cadmium	62	48	90349	Zirconium	50	40	88764	671	keV
Magnesium	12	12	13930	Silicon	14	14	21491	Neon	10	10	7043	674	keV
Gadolinium	93	64	70825	Erbium	98	68	64921	Neodymium	88	60	77407	678	keV
Aluminum	14	13	17194	Silicon	15	14	21893	Magnesium	13	12	13190	695	keV
Antimony	70	51	89588	Xenon	74	54	89861	Cadmium	66	48	90019	704	keV
Palladium	62	46	89523	Tin	68	50	91653	Molybdenum	56	42	88115	722	keV
Samarium	88	62	77049	Gadolinium	93	64	70825	Neodymium	83	60	84001	728	keV
Osmium	114	76	38699	Mercury	121	80	27672	Hafnium	107	72	50462	736	keV
Barium	81	56	87733	Cerium	82	58	88081	Xenon	80	54	88125	740	keV
Osmium	112	76	41125	Platinum	116	78	34765	Tungsten	108	74	48228	743	keV
Thallium	122	81	25988	Lead	126	82	21758	Mercury	118	80	30964	746	keV
Osmium	114	76	38699	Platinum	120	78	29921	Tungsten	108	74	48228	751	keV
Platinum	118	78	32652	Mercury	122	80	27356	Osmium	114	76	38699	751	keV
Zirconium	52	40	84822	Molybdenum	55	42	85544	Strontium	49	38	84868	768	keV
Selenium	43	34	74601	Selenium	46	34	77761	Selenium	40	34	72212	771	keV
Gadolinium	92	64	72536	Erbium	98	68	64921	Neodymium	86	60	80923	772	keV
Tellurium	70	52	90304	Xenon	74	54	89861	Tin	66	50	91526	779	keV
Gallium	38	31	69321	Germanium	42	32	73422	Zinc	34	30	66001	781	keV
Strontium	46	38	80641	Molybdenum	50	42	86807	Selenium	42	34	75259	784	keV
Krypton	48	36	82431	Strontium	50	38	87910	Selenium	46	34	77761	809	keV
Hafnium	107	72	50462	Tungsten	112	74	42498	Ytterbium	102	70	59250	824	keV

Table 2. (Continued)

1	2	3	4	5	6	7	8	9	10	11	12	13	14
Zinc	37	30	67879	Gallium	38	31	69321	Copper	36	29	67261	824	keV
Osmium	113	76	38978	Platinum	114	78	36283	Tungsten	112	74	42498	825	keV
Krypton	47	36	79984	Strontium	51	38	86203	Selenium	43	34	74601	836	keV
Silver	62	47	88722	Palladium	64	46	88335	Cadmium	60	48	89951	842	keV
Silver	60	47	88404	Tin	65	50	90035	Ruthenium	55	44	87619	846	keV
Erbium	99	68	63286	Ytterbium	102	70	59250	Dysprosium	96	66	68181	859	keV
Tungsten	109	74	46347	Osmium	112	76	41125	Hafnium	106	72	52434	865	keV
Antimony	72	51	89217	Xenon	78	54	89286	Cadmium	66	48	90019	871	keV
Technetium	54	43	87224	Ruthenium	55	44	87619	Molybdenum	53	42	87712	883	keV
Tellurium	73	52	89019	Barium	80	56	88906	Cadmium	66	48	90019	887	keV
Palladium	59	46	88422	Tin	65	50	90035	Molybdenum	53	42	87712	903	keV
Silver	62	47	88722	Tin	67	50	90398	Ruthenium	57	44	87951	905	keV
Tellurium	68	52	89404	Xenon	72	54	89162	Tin	64	50	90561	915	keV
Rhodium	58	45	88024	Cadmium	63	48	89254	Molybdenum	53	42	87712	918	keV
Germanium	42	32	73422	Selenium	46	34	77761	Zinc	38	30	70006	923	keV
Krypton	44	36	77897	Strontium	48	38	84512	Selenium	40	34	72212	930	keV
Neon	12	10	8026	Magnesium	14	12	16212	Oxygen	10	8	783	943	keV
Iridium	114	77	36698	Mercury	116	80	31846	Tungsten	112	74	42498	948	keV
Silver	60	47	88404	Tin	62	50	88658	Ruthenium	58	44	89100	950	keV
Indium	64	49	89372	Tellurium	70	52	90304	Palladium	58	46	89400	960	keV
Mercury	119	80	29557	Lead	124	82	23795	Platinum	114	78	36283	964	keV
Platinum	118	78	32652	Lead	124	82	23795	Tungsten	112	74	42498	989	keV
Oxygen	9	8	809	Neon	11	10	5733	Carbon	7	6	-3125	990	keV
Titanium	24	22	44122	Nickel	30	28	60224	Sulfur	18	16	29013	993	keV
Neodymium	86	60	80923	Samarium	90	62	74761	Cerium	82	58	88081	996	keV
Selenium	43	34	74601	Strontium	46	38	80641	Zinc	40	30	69559	998	keV

In chemistry, the aforementioned reaction of obtaining chlorine and hydrogen from hydrochloric acid can serve as an analogue of such processes. In this case, the analogue of the nucleus, consisting of particles in the form of protons and neutrons, is a molecule consisting of particles in the form of atoms. This reaction can be represented as follows. Decomposition of a molecule HCl into atoms H and Cl (Fission). This consumes some energy. Then the union into other molecules of two chlorine atoms and two hydrogen atoms (Fusion). This releases some energy. The resulting input energy is equal to the difference between the released and expended energies.

However, this process can be presented in a different way. That positive energy that arises when atoms combine into a new molecule takes part in the process of separating an atom from another molecule. Therefore, when two molecules interact, less energy is required to separate an atom from the interacting molecules than when the molecules are isolated from each other. The same situation occurs with nuclei and the particles of which they are composed.

As follows from the tables presented, the average value of the excess energy is smaller 100 keV or about 10^{-12} J. This is approximately a four orders of magnitude greater than the excess energy in exothermic chemical reactions, and approximately four orders of magnitude smaller than the energy in the decay of uranium nuclei. As is seen, there are nuclear reactions where the excess energy is about few tens of keV.

The table contains a lot of interesting information that requires separate detailed comments. For example, in accordance with string where the excess energy is equal 990 keV, nucleus of oxygen losses alpha particle and is transformed into nucleus of carbon. From this and similar information we can conclude that the presented approach is highly probable.

The question arises - which elements are subject to nuclear transformation in cavitation vortex generators? One liter of water contains the following substances (in micrograms), which can be drawn into the cavitation bubble and participate in nuclear transmutations: {Al, 500}, {Ba, 300 }, {Be, 500 }, {Cd, 50000 }, {K, 180 000 }, {Si, 10 000 }, {Mg, 40 000 }, {Mn, 100 }, {Cu , 1000 }, {As 50 }, {Ni, 100 }, {Pb, 30}, {Sr, 7000 }, {Cl, 35 000}, {Cr, 50 }, {Zn, 50 00}, {Mo, 250 }. As follows from the table, there are many candidates for this purpose.

We tried to show that when using the wave approach to the consideration of the properties of particles of the micro world, a situation is possible in which a strong interaction between nuclei can take place without overcoming the Coulomb barrier. A decrease in the magnitude of an interaction between nuclei in a form of corpuscles can be compensated by an increase in the interaction time if wave properties of nuclei are used and they are considered as oscillators. For oscillatory systems, an increase in the interaction time can lead to the same result as an increase in the interaction magnitude for corpuscles.

Nuclear physics experts deny the possibility of nuclear reactions without breaking the Coulomb barrier. This belief is based on a corpuscular approach to particles of the micro world. However, they cannot object to the fact that particles have wave properties, since this fact has irrefutable experimental evidence. They should come to terms with the fact that the interaction between nuclei can differ from that which exists between the particles and stop unjustified criticism of the possibility of implementing nuclear transformations. Moreover, at present this possibility has numerous experimental confirmations.

From the presented consideration, we can conclude that a significant interaction between neighboring nuclei is random, and depends on many parameters that characterize the nuclei. From here follows a simple conclusion. It is necessary to bring the nuclei as close as possible to the minimum possible distance at a relatively long time. Such an opportunity can only be realized using the light, since the time constant of light processes is 6 orders of magnitude less than that for mechanical displacements. For maximum convergence of the nuclei, maximum pressure is required. This pressure is produced by the circulating light. The circulating light opens up a new world in which nuclear transmutations can be implemented.

Further, in order to increase the energy output, the described procedure should be repeated as often as possible. In accordance with this requirement, the following conditions must be met. Firstly, the time to execute one procedure must be minimal and, secondly, the procedure must be performed simultaneously in many different regions of space. These conditions are met when using the circulating light.

Conclusion

We tried to find an area where the unique physical parameters of the circulating light could be useful. A clue in this direction was that a combination of words "excess energy" was often found in the texts regarding luminous anomalous objects. Bearing in mind the law of conservation of energy, we assumed that excess energy can be obtained from nuclei, but not with the help of well-known methods Fission and Fusion, but by extracting nuclear energy from all kinds of nuclei during their mutual interaction. Bearing in mind the wave properties of nuclei and considering them as oscillators, we came to the conclusion that interaction is fundamentally possible between identical distant nuclei, if their natural frequencies are the same. The result of the interaction can be as large as possible, provided that the positions of the nuclei in space remains unchanged in time for a relatively great period of time.

This condition can only be met by chance if the gas of nuclei, whose phases are randomized as a result of numerous short-term collisions of atoms, is sharply compressed to such an extent that the atoms are densely packed and the nuclei in neighboring atoms can interact with each other for a relatively long time. At the same time, in some regions situations may occasionally occur where the phases of the oscillations are favorable for the transfer of energy from one nucleus to another. With such a transfer, a situation is possible in which one of the nuclei turns out to be unstable, and a nuclear reaction occurs with the release of energy. To obtain a significant output of energy, it is necessary to perform this operation as often as possible.

The circulating light is the most suitable for solving this problem. Firstly, the circulating light carries out the ultimate compression of matter, at which atoms are densely packed (evidence of this is the record lifetime of light in such a substance, which is thousands of times greater than the lifetime of light in the Earth's atmosphere). Since the light is scattered by inhomogeneities in the density of matter arising as a result of the Brownian motion of molecules, then such a long lifetime indicates that the Brownian motion is reduced to a minimum). Secondly, such compression is carried out extremely quickly, since transient processes are determined by the light speed. In addition, compression is carried out in a thin film, which requires a displacement of atoms over minimal distances.

To the greatest extent, these requirements are met by bubbles in the water that have arisen as a result of the cavitation or the introduction of air bubbles into the water. This conclusion has experimental confirmation. Commercially available thermal energy generators and operating devices from the Rosch Innovation firm, which show excess energy, are known.

Chapter 4

Circulating Light and Mechanical Excess Energy

What we know is a drop, what we don't know is an ocean.
I. Newton

Introduction

Having studied the properties of the circulating light, knowing its unique parameters, and bearing in mind the above analysis of possible nuclear transformations under the action of the circulating light, it would be unreasonable to ignore numerous attempts to develop devices that extract the excess energy from nuclear reactions. In principle, the availability of excess energy makes it possible to speak of the creation of a fuel-free engine, if the excess energy is sufficient to compensate for the inevitable energy losses during the operation of any device in which mechanical movements take place.

Currently, there are thousands of projects on the Internet to create fuel-free devices. A number of patents for such devices is also measured in hundreds. However, the situation is much worse with the video, which demonstrates the operation of these devices. The situation is even worse with the video, which demonstrates the production of useful energy using such fuel-free devices.

Recently, sources of heat energy have been developed in which heat is obtained from electricity with high efficiency. In this case, unlike traditional heating devices using resistors through which electric current is passed, electric motors are used that generate heat by using the phenomenon of cavitation. It is shown that such devices generate excess heat, violating the well-known Joule law. This is confirmed by the existence of quite a few commercially available devices that use the phenomenon of cavitation to effectively heat water. Only in Russia, hundreds of installations with a power from units to hundreds of kilowatts operated. Currently, such heat generators are already mass-produced. We present as an example only one item from a great list of organizations only in Russia that supply heat generators to the market.

OOO Ecoteplo Russia Moscow, Volgogradsky prospekt, 42 installations, maintenance, supply of equipment warranty service.

"Vortex heat generators" (VHG) are modern, highly efficient, autonomous, energy-saving, environmentally friendly heating, heat supply and hot water systems. Heating costs are lower than when using central heating by 3-5 times, heating and electrode boilers by 3-5 times, diesel by 8-10 times, gas by 15%. Serially produced, certified VHG are a standard asynchronous electric motor 3000 rpm, supply voltage 380 V, mounted on the same frame with a heat generator that converts mechanical energy into thermal energy. They are fully prepared for connection to a new or existing heating system, and the design and dimensions of the VHG simplify its placement and installation in a heating unit. The equipment does not require approval from RosTechNadzor, it is supplied starting from the heating season 2004/2005. Currently, more than four hundred VHGs are operated in the regions of the Russian Federation, near and far abroad.

Obtaining excess energy due to cavitation phenomena is proposed in many patents. Here is some of them.

Table 3. List of patent about excess heat energy production by means of cavitation

US patent	**4333796**	**1978**	**H.G. Flynn**	**Energy generation method using acoustic cavitation**
US patent	424 797	1981	Y. Perkins and P. Poup	Heating device
US patent	5188090	1991	Griggs J. L.	Apparatus for heating liquid
USSR patent	2054604	1996	A, F. Kladov	The method to get energy
RU patent	2165054	2000	Potapov Yu. C	The method to get heat
RU patent	2156412	2000	V. N. Kirilenko, C. O. Brulev	Method for hydrodynamic heating liquid

Thus, the excess heat that occurs during cavitation is an objective reality, although there is currently no generally accepted scientific explanation for this phenomenon.

A fundamentally new step was the development of devices in which excess energy is converted not into heat energy, but into electrical one. Known methods of converting thermal energy into mechanical one used in steam locomotives and thermal turbines are not applicable in this case, since the temperature difference between the heater and the refrigerator is small, which gives an unacceptably low efficiency.

As far as we know, the first device capable of operating without external power sources was publicly demonstrated in 2014 by Rosch Innovation firm. In this case, part of the energy generated by the device can be fed into the power grid. As a result, this device is a fuel-free source of energy. This means that the device does not need traditional energy sources such as oil, gas, nuclear, solar, hydro and wind power to generate electricity. Later, similar devices were developed and demonstrated in many countries such as India, Thailand, Taiwan, Indonesia, New Zealand. Since these devices do function without consuming energy from external sources, their operation should be carefully studied and analyzed to determine whether the exothermic nuclear reactions, the possibility of existence of which in the world of the circulating light was discussed above, are the source of energy.

Device of the Rosch Innovation firm

The German company Rosch Innovation showed a device in which two gears connected by a chain drive rotate indefinitely. Since then, a number of skeptics that this cannot be, because the law of conservation of energy is violated, has not diminished. Although the Rosch Innovation firm advertises power plants with a capacity of 500 and 5000 kW, no single energy source has been built in Germany since time for energy production.

We believe that Rosch Innovation really demonstrates a perpetual motion machine that works without the consumption of well-known and generally accepted types of energy. To show that this is possible, let us first analyze the video of a true working perpetual motion machine.

It is quite simple to describe its construction. Nine identical barrels 1 attached to a chain 2 at the same distance between them as is shown in Figure 51. The chain is put on two gears 3 that can rotate around their axes. On the other side of the chain, the same number of barrels are attached, so that such a system has the position of indifferent equilibrium at a steady-state like two identical weights suspended from two identical ends of a weightless thread thrown through a block. It can be clearly seen from this figure that the same barrels are attached on the other side of the closed chain. All barrels are immersed in a container 4 of water. To put the device into operation, convention air is blown into each lowest barrel on the right side of the chain.

Besides, there is the following additional information. Compressed air is blown into the opening in the lowest barrel from the compressor 5. The compressor is set in motion through a system of gears connected to the gear 3 through which the chain is thrown. It must be assumed that the energy that the compressor takes from the device, as well as the energy spent on water resistance and friction in the elements of the chain and gears, is compensated by other energy produced in the system.

Subsequently, the installation was improved in such a way that the barrels were replaced by half barrels (baskets), as shown in Figure 51b.

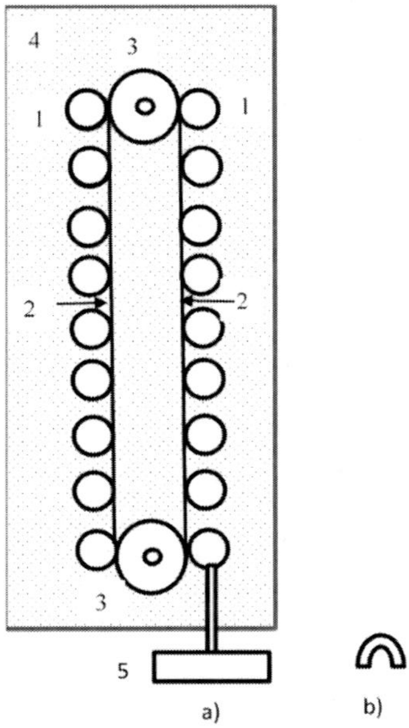

Figure 51. Rosch Innovation device diagram. a) total scheme; b) form of basket.

Consider first the reasons that make the system move. The weight of the blown air is negligible. In the same time there is the Archimedes force F applied to each right basket. For given case, in accordance with the Archimedes law, the force is equal to the weight of the water, the volume of which is equal to the volume of inserted air. Thus, the equilibrium is violated and the right buckets pull up chain at force $9F$. We can estimate the velocity of baskets from demonstration video. The chain moves at a speed about of 0.5 m/s. If volume of 1 liter is inserted in each basket, then the force with which the basket pulls up the chain is 10 Newton. Nine baskets pull up the chain with force 90 Newton. Then the power of the entire device is equal to the product of the force and speed, that is, equal to 45 W. That is why till now we have not seen a single source of energy based on this perpetual motion machine. Nevertheless, for many years, advertising of these devices has continued with the promise of creating installations from several kilowatts to several megawatts.

Let us estimate the chain tension forces in the setup of 1 MW power. The speed of the chain is known and is equal to about 0.5 m/s. Since the power is equal to the product of the force and the speed, we can calculate the third term from two known terms. As a result, we get that the force is 2 Mega Newton or 200,000 kilograms of force or 200 tons. What should be the dimensions of the support to hold such tension? This is the weight of about 200 cars. Even the chain tension is 1000 times less, that is, a tension of 200 kg of force will crush the demonstrated installation, the diagram of which is shown in Figure 51. The Germans would not refuse free energy if it were available with the help of such installations. Possibly, the authors of the perpetual motion machine were not fully aware of the physical phenomena responsible for its operation. Indeed, those are right who argue that suckers are not mammoths, they will never die out.

It remains to find out where the excess energy comes from, which compensates for the inevitable energy losses. The answer is very simple. Energy is supplied by means of great compression of atoms of gases dissolved in water. The compression is produced by air bubbles inserted in water. The bubbles are generated by compressor 5 that injects the air in water. We showed early that the excess energy can be obtained in bubbles arising due to cavitation. These bubbles are filled by water vapors and gases dissolved in water. The temperature of bubbles increases because of compression of gases with a decrease of a diameter of the air bubbles. The great compression is produced due to the surface tension of water since the curvature of the air balls is very great. It is may to assume that these processes and processes that take place in the cavitation bubbles at phenomenon of sonoluminescence are identical. This phenomenon of sonoluminescence has been investigated in detail because the temperature of the cavitation ball measured by spectroscopic method is about 100 000^0 C.

Careful studies of this phenomenon with the publication of detailed experimental results were associated with the hope of using this phenomenon to carry out a thermonuclear fusion reaction. We have shown that the shift of the emitted light to the short-wavelength region of the spectrum, which was accepted by researchers as the radiation of a body with an extremely high temperature, has a different explanation. The shift is explained by the decrease in the wavelength of the circulating light as the ball size decreases. The presence of the circulating light also explains other observed anomalous effects. Thus, the presence of the circulating light in the ball is proved in explaining anomalous experimental results (the presence of circulating light in Ball Lightning is also proved by explaining its observed anomalous properties).

Where does the circulating light come from in the ball? The explanation of the appearance of the light that the ball emits is not given by us, but by the researchers of the phenomenon of sonoluminescence. This is due to an appearance of a shock wave during a rapid reduction in the size of the ball. At the front of the shock wave, the temperature rises to several thousand degrees. Heated atoms emit light in all directions. Part of this light begins to circulate in the shell of the compressed gas and is accumulated gradually (exactly the same explanation is given in the analysis of the processes leading to the occurrence of Ball Lightning with the help of a shock wave that occurs during the discharge of ordinary lightning [11]).

It can be expected that the same processes occur in air bubbles that are formed when air is introduced into water. The difference between air bubbles and cavitation bubbles is that the air bubbles are filled with gas immediately upon their occurrence. The cavitation bubbles after their appearance are filled initially with water vapor and gases dissolved in water. When the bubble is compressed, water vapor condenses, and gases remain. Evidence that gases play an essential role in the origin of light is the fact that the dissolution of inert gases in water increases the light radiation by hundreds of times. This phenomenon is also explained [47].

Of course, the introduction of the air must be carried out through small holes under high pressure, so that balls of a minimum diameter can be obtained in the water. In this case, the surface tension of the water creates the very great air pressure. This pressure tends to compress the balls to such an extent that the temperature of the air rises and the circulating light appears inside the surface of the small ball.

We have previously presented arguments, both theoretical and experimental, that circulating light creates the conditions for the emergence of excess energy as a result of nuclear substitution reactions. An increase in the temperature of the shell in which the light circulates leads to an increase in the temperature of the air in the entire bubble, which, in turn, leads to an increase in the volume of the bubble.

The increase in the volume is accompanied by positive work, which is done by the forces arising from the increase in the volume of the bubble. As a result, the energy stored in the mixture of water and bubbles in the gravitational field of the Earth increases. Thus, the resulting thermal energy is converted into the potential energy of a mixture with reduced density, surrounded by water with the original density in the gravitational field of the Earth. The Roche installation converts this potential energy into mechanical energy, which keeps the compressor running.

In the following, we will refer to the described effect of an excessive increase in the volume of a liquid with air bubbles due to nuclear reactions as the *Roch effect.*

This is not the place to explain the theory of the emergence of excess energy in cavitation bubbles. It is sufficient to refer to the fact that only in Russia there are more than 10 companies that offer heating elements that emit heat, the energy of which is noticeably greater than the consumed electrical energy. The reason for the appearance of the excess energy is nuclear transformations in the bubbles that arise in water. In the heating elements, such bubbles arise as a result of cavitation. In the considered case, they are simply introduced into the water by means of a compressor.

In all the mentioned heating setups, the excess energy is converted into thermal energy, which heats the water. The coefficients of the thermal expansion of water depend on its temperature. At a temperature of 10-30°C, the coefficients of the thermal expansion of water are about 20 times less than those of gases. Therefore, heating the gases inside the bubble to reduce the density of the liquid has an effect 20 times greater than heating the water. It is this circumstance that is used in the installation of the Rosch Innovation firm. In addition, unlike water, gases can be heated to much higher temperatures.

The release of heat as a result of nuclear transformation leads to a significant increase in the temperature of the gases inside the bubble, which entails an increase in the volume of the gas. For example, an increase in temperature by 300^0 C leads to an increase in volume by 2 times, and an increase in temperature by 3000^0 C leads to an increase in volume by 10 times. With an increase in volume, the Archimedes force increases accordingly, and, consequently, the total power of the setup increases also.

The described processes can be compared with those in an internal combustion engine. In this case, the analogue of the bubble of variable volume is a cylinder of variable volume due to a movable piston. With an increase in the temperature of the combustible mixture during its combustion heat releases, the temperature of the mixture increases that entails an increase of the mixture pressure. As a result, the volume of the cylinder increases since the pressure pushes the piston. In this case, the gases perform mechanical work, which is converted into the kinetic energy of the flywheel, and then into mechanical work, ensuring rotation of wheels.

It can be said that in the installation under consideration there are millions of miniature "cylinders with a piston" in the form of bubbles, and the heating of gases occurs not due to the combustion of a combustible mixture, but due to the combustion of some unknown elements contained in the air bubble because of nuclear reaction. This increases the volume of the "cylinder with piston." However, in this case it is not possible to transfer this mechanical movement of the numerous miniature bubbles to the direct rotation of the flywheel.

The firm Rosch Innovation solved the problem of combining the many emerging energies in millions of air bubbles into the energy of rotating gears as follows. Water with bubbles is placed in a container. As the volume of bubbles increases, the density of water decreases. Further, the circumstance that such a system is located in the gravitational field of the Earth is used. If such a container is placed in water and the weight of the container is balanced with the help of a block and a rope, at the other end of which is attached exactly the same container filled with ordinary water, then the balance will be disturbed, since the weight of the container with water of reduced density is smaller than that of the same container filled with ordinary water. As a result, a force equal to the difference in the weights of these containers acts on the rope.

Thus, miniature "cylinders with a piston" are continuously created in the installation under consideration by blowing air. Unlike a conventional combustible mixture, the temperature of which rises due to a chemical oxidation reaction, the temperature of air in air bubbles rises due to the nuclear transformation of unknown elements present in air or water. In both cases, before combustion, the combustible mixture is subjected to strong compression.

We can say that the device of the Rosch Innovation firm is a heat machine that converts the heat energy into mechanical one. Unlike traditional heat engines such as an internal combustion engine, a turbine engine, the Stirling engine, the device under consideration additionally uses gravity, which sums up a lot of mechanical energy that has arisen from heat energy in each of a million air bubbles into a single mechanical energy that drives rotating gears.

It is becoming clear why the firm has moved from using barrels to using baskets. In barrels, the water temperature gradually increases as no water is poured out of the barrels. It is known that the water surface tension responsible for compression of air within the bubble decreases sharply at the water temperature of more than 40^0 C. The use of baskets from which water is completely poured out eliminates this disadvantage.

It is also known the property that the release of excess energy in water, which is not renewed, gradually weakens in time. Presumably, this is due to the fact that the unknown for us elements that cavitation bubbles are using to extract energy gradually exhausted. In the baskets, water is constantly renewed and this effect is absent.

This setup simply draws additional energy from air bubbles where additional energy is produced as a result of nuclear transformation. In the proposed version, such a setup is simply a useless toy, since the costs of its production do not pay off with the energy it generates. The main drawback of this setup is the low mechanical speed, which leads to its bulkiness. Why are collector motors used in hand-held electric tools, although conventional asynchronous motors are much easier to manufacture and operate? The explanation is simple. The speed of rotation of an asynchronous motor cannot be more than 3000 rpm, and the rotational speed of a collector motor can be 10 times higher. As a result, its weight is significantly reduced, and therefore the convenience of working with it increases. Why are Japanese car engines lighter and more compact than American ones? Because they have a rotation speed 2 times higher.

The method used in the setup to extract excess energy from air bubbles can be used much more efficiently to get mechanical energy. Below we present several devices for which applications for Russian patents have been filed. The applications were filed for the sole purpose of preventing other inventors from filing similar patents.

Improved Devices

Since devices of the Rosch Innovation firm operates without external sources of energy, we need to conclude that excess energy is not a myth but an objective reality. Taking into account above analysis, we present several modifications of the device, in which the same principle of obtaining excess energy are used.

We suppose that the excess energy is produced by air bubbles, that produce the circulating light. As was shown the circulating light is favorable for an occurrence of nuclear transformations. The basis for this conclusion is numerous extensive results of studying the phenomenon of sonoluminescence, in which an anomalous glow of the cavitation bubble was investigated in detail. The explanation of these anomalies on the assumption that trhe circulating light is responsible for their occurrence gives reason to believe that the circulating light also occurs in other bubbles obtained in a different way. In particular, the presence of the circulating light in the bubbles that occur during hydrodynamic cavitation makes it possible to explain the phenomenon of cavitation corrosion of metal objects.

Below are some fantasies about how water can be used to extract energy, the shortage of which is felt throughout the world. Devices belong to the world of the circulating light, since in our opinion it is circulating light that allows these devices to be put into action.

Preliminary Remarks

To understand the operation of the devices discussed below, some elementary information from physics and an understanding of the meaning of the terms used below are required. These terms, such as speed, pressure, mass, force, volume and density of water, torque are known from a high school physics course.

Since these concepts are applied to the flow of water, some notions need to be clarified to this particular case. The momentum of some object is defined as the product of its mass and its velocity.

The water consumption is understood as the mass of water that flows through the cross section of a pipeline in one second.

The flow momentum of water in a pipeline is defined as the product of the water consumption and its velocity.

Instead of a compressor that produces air to pump into the basket used by the Rosch Innovation firm, we propose to use a simple device called the Venturi tube. The design of this device is shown in Figure 52.

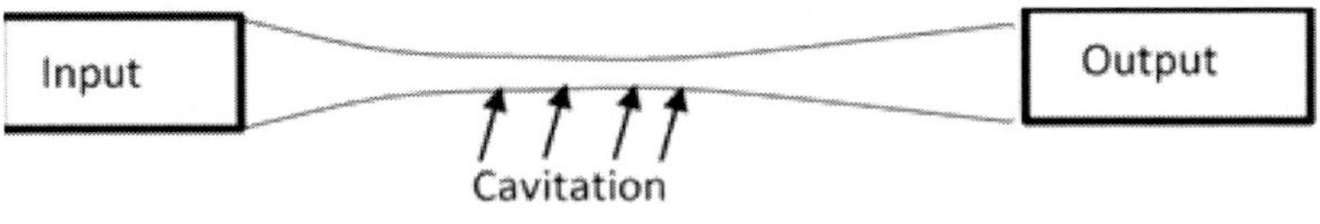

Figure 52. Design of the Venturi tube.

The Venturi tube is inserted into a cut of a conventional pipe through which water flows. In a narrow section of the Venturi tube, the pressure can drop to such an extent that water boils at a low temperature. When water boils, cavities are formed that can be filled with gases dissolved in water or gas coming from the atmosphere through a special tube without any compressor because the atmospheric pressure exceeds the pressure inside the Venturi tube. This use of the Venturi tube is not new and is used in known devices called ejectors.

To increase the efficiency of the Venturi tube, on the side of the tube in the part where its diameter is minimal, small holes are made through which ambient air can penetrate into the water flowing through the Venturi tube. As a result, a mixture of water and air bubbles is created in the flowing water without the use of the compressor, which requires energy to operate. Just as in the Roch installation, the Roch effect occurs, in which the volume of the mixture increases to a greater extent than follows from the law of conservation of energy.

After passing through the Venturi tube, the density of water decreases due to the Rosch effect. However, the masses of water passing through the inlet and outlet of the Venturi tube are the same. Therefore, the volumes of water passing through the inlet and outlet of the Venturi tube should be different. The volume of water passing through the outlet is greater than through the inlet. This may be the case if the speed of the water passing through the outlet is greater than through the inlet.

Thus, the water consumptions that passes through inlet and outlet are the same but velocity of water in the outlet is greater than that in the inlet. In this case, the momentum of water at the outlet of the Venturi tube is greater than at its inlet. Water increases its momentum as it passes through the Venturi tube. For example, if 100 liters passes through the Venturi tube per second and the speed of the water increases by 10% from 10 m/s to 11 m/s, then the momentum of the water increases by 10 kg m/s^2. According to Newton's second law the force applied to the water inside the Venturi tube is equal to 10 Newton or 1 kg acts on the water. The force is directed in the direction of water movement. Thus, the Venturi tube makes the water flow faster.

Since the water consumption at the inlet and outlet of the Venturi are the same, the products of the density of water and its velocity at the inlet and outlet of the Venturi are the same also. At the same time, the velocity of water at the outlet of the Venturi tube is greater than at the inlet. Therefore, the product of the density of water and the square of the velocity at the outlet of the Venturi tube is greater than at its inlet. This product is equal to the kinetic energy of water. Therefore, we can state that the Venturi increases the kinetic energy water, that is, it is a source of energy. In this case, the amount of generated energy is proportional to the product of two parameters. The first parameter γ

is equal to the relative increase in the velocity of water flowing through the outlet of the Venturi tube compared to the velocity of the water flowing through the inlet of the Venturi tube. The second parameter is the water velocity v at the entrance to the Venturi tube.

To increase the power delivered by the Venturi tube, both parameters must be increased. The value of the first parameter is unknown to us and we can only refer to the operating setup of the Rosch Innovation firm. As for the second parameter, this parameter can be chosen as much as possible. For example, if the water speed is increased to only 10 m/s, then the efficiency of the setup will increase by 20 times compared to the setup of the Rosch Innovation firm, where the water movement speed is 0.5 m/s.

Thus, assuming that the density of water at the outlet of the Venturi due to the Rosch effect decreases by a factor of $(1 + \beta)$ compared to the density of water at the inlet, we have the following effects

The speed of water increases by $(1 + \beta)$ times.

The momentum flow of water at the outlet of the Venturi tube increases by $(1 + \beta)$ times.

The kinetic energy of water at the outlet is greater than at the inlet by $(1 + \beta)$ times.

The Venturi tube is an analogue of a compressor that increases the energy of the water flow by $(1 + \beta)$ times.

According to the estimations of the processes in the basket of the Rosch Innovation firm, the value of β is in the range of 0.2 - 0.5 (ratings are based on real work, not advertising). The same value can be obtained from the analysis of the average increase in thermal energy in vortex heat generators.

Gravity Device

Knowing the physical phenomena that ensure the operation of the device of the Rosch Innovation firm, let us show that this device can be improved significantly, getting rid of almost all moving parts, leaving only a generator that generates electricity.

With these considerations in mind, an improved version of the Rosch Innovation device is shown in Figure 53a.

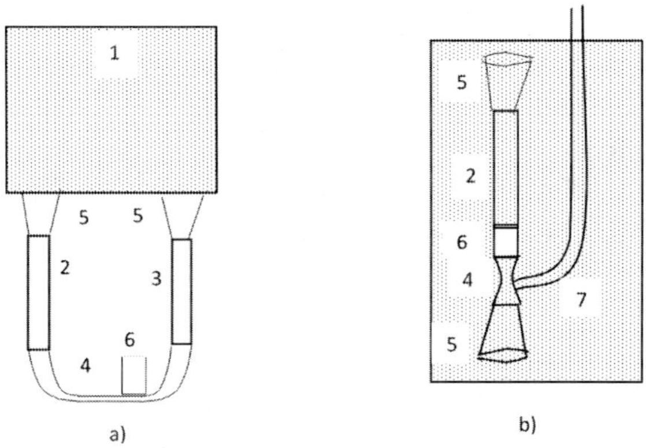

Figure 53. A source of heat and mechanical energy. a) simplest scheme; b) improved scheme completely submerged in water.

Here two vertical tubes 2 and 3 are connected by means of curve tube 4 with the Venture tube inside. The tubes are inserted in container of water 1. Funnels 5 are added to reduce losses during the input and output of water into the tubes. A centrifugal pump 6 is inserted inside the pipeline, consisting of pipes 2, 3 and 4. Pump 6 is attached outside the Venturi tube. The purpose of this pump is to, at the initial stage, when the device is brought to normal mode, to give the water flowing through the narrow part of the Venturi tube, such a speed at which the Roch effect occurs in the tube. At the same time, unlike the Rosch Innovation device, there is no need to use a compressor for these purposes.

The device works as follows. When the device is started, the centrifugal pump 6 pumps water from pipe 3 to pipe 2 through the Venturi tube 4. The water pressure decreases in the Venturi tube and the Rosch effect occurs. As a result, the density of water in pipe 2 becomes smaller than that in pipe 3 and the water pressure ΔP between the inlet and outlet of the Venturi tube arises because the weight of water in the right tube 3 is greater than that in the left tube 2. If this water pressure ΔP is sufficient to provide a flow water through the Venturi tube at speed that causes the Rosch effect and provides pressure ΔP, then the pump 6 can be turned off.

Water will circulate due to the resulting pressure difference ΔP in pipes 3 and 2. At the same time, passing through the disconnected from source of electrical energy centrifugal pump 6, the water will rotate its impeller and the centrifugal pump will operate as a turbine. It is known that electrical machines have the property of reversibility. When voltage is applied to them, they work as an electric motor. If the shaft of the electric motor is rotated by another engine, then the electric machine performs the function of a generator of electric energy. Thus, pump 6 performs various functions in different time. It operates as a centrifugal pump when the device is put into operation and it is required to provide a pressure difference in pipes 3 and 2. It operates as an electricity generator in stationary mode, when the pressure difference in pipes 3 and 2 is maintained due to the Rosch effect in the Venturi tube.

Of course, this device cannot operate without additional energy, which must compensate for the inevitable losses that occur during the movement of water in the pipeline, as well as when water enters and exits the pipeline. For the operation of the device, it is necessary that the resulting additional energy due to nuclear transformations be greater than the energy losses in the device. The fact that additional energy can be obtained follows from the successful operation of the Rosch setup and cavitation heat generators.

The total losses in the system can be reduced by almost 2 times, and the design of the device can be simplified if the device is completely submerged in water. In this case, the need for pipe 3 is eliminated. We can assume that the pipe 3 has a very large diameter. In this case, the pressure losses that were in the pipe 3 disappear and the efficiency of the device increases. The design of such a device is shown in Figure 53b. In this case, additional tube 7 is required to take atmospheric air into the Venturi tube.

Thus, when analyzing the forces in Figure 53, we must keep in mind three phenomena that are responsible for the positive pressure in the pipeline. This is the difference in the densities of water in tubes 2 and 3 due to the Rosch effect. In the field of gravity, this leads to a pressure difference in tubes 2 and 3 and an appearance of a positive pressure. Let's call it the *gravitational effect.* The same effect is used in the setup of the Rosch Innovation firm. Under weightless conditions, this setup cannot work.

The second phenomenon, which leads to the appearance of an additional positive pressure, is associated with an increase in the momentum of water at the outlet of the Venturi tube relative to its inlet. Let's call this phenomenon the *momentum effect*. This effect is analogous to the operation of a compressor, which increases the outlet water pressure compared to the inlet pressure. However, unlike a conventional compressor, it is not the pressure that increases at the outlet, but the speed of the water movement. Nevertheless, the water flow at the inlet and outlet of the compressor is the same. The speed is increased due to a corresponding decrease in the density of the water.

The third effect is associated with an increase in the kinetic energy of water flowing through a Venturi tube due to nuclear transformations in air bubbles. Let's call this phenomenon the *excess energy effect*. In the case under consideration, all three effects owe their origin to an increase in the volume of air bubbles due to the heating of the gas contained in them during nuclear transformations. The first effect is manifested under the additional condition that the vertical tubes are located in Earth's gravity field.

For completeness, let's mention a fourth effect, which occurs in the field of centrifugal forces, which can replace gravitational forces in alternative setups. We call this phenomenon the *centrifugal effect*, the application of which is discussed below.

It should be borne in mind that with an increase in the depth of immersion of the Venturi tube, the water velocity in the Venturi tube, at which water pressure ensures the occurrence of the Rosch effect, increases. The dependence of the required water velocity on the height of the pipeline is shown in Figure 54.

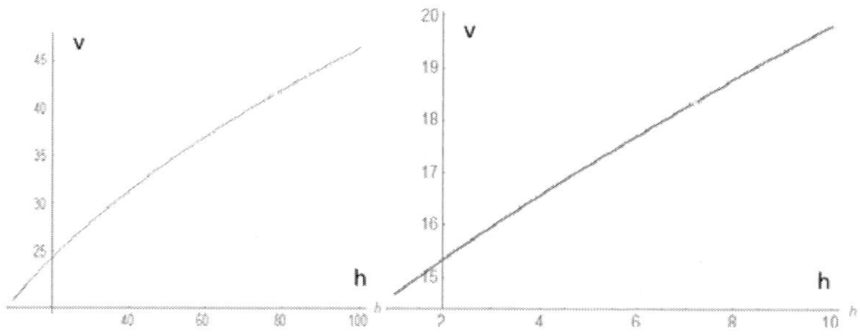

Figure 54. Dependence of the velocity in the Venturi tube that is required for arising cavitation on the height h of the pipeline.

As is seen, an increase in the height of the pipeline from 10 m to 100 m, the water velocity should be increased from 20 to 45 m/s. The required speed in the Venturi tube is achieved by selecting its appropriate diameter in its narrow part. In this case, it should be taken into account that the resistance of the Venturi tube increases with an increase in the ratio of the diameters of the pipeline and the diameter of the narrow part of the Venturi tube. In addition, it should be borne in mind that the pressure drop in the pipeline when water flows through it is proportional to the square of the water velocity. Unfortunately, the book is devoted to the circulating light, and not to the design of devices based on the use of nuclear energy, so we will limit ourselves to the above considerations.

It is easy to see that the presented devices are so simple that their dimensions can vary widely. At the same time, it should be borne in mind that the losses are proportional to the length of the pipeline and inversely proportional to its diameter. The positive effect in a form of an increase of a difference of pressures in vertical tubes is also proportional to the length of these tubes. Thus, one of the ways to reduce losses is to increase the diameter of the pipeline. In this case, the power of the device increases in proportion to the square of the diameter and the length (height) of the pipeline. Losses in the pipeline can be reduced also with a decrease in the roughness of the inner surface of the pipeline.

At the present time we do not know by what percentage the water density can be reduced after the passage of water through the Venturi tube. However, it is well known about the magnitude of losses at various parameters of the pipeline. The graphs in Figures 55–58 show how the various parameters of the device depend on the diameter d of the pipeline used and its length h. Each figure is represented by two graphs a) and b). The graphs differ in the value of the parameter β introduced in the previous section. In this case, the percentage of inevitable energy losses in the entire system must be subtracted from the parameter β.

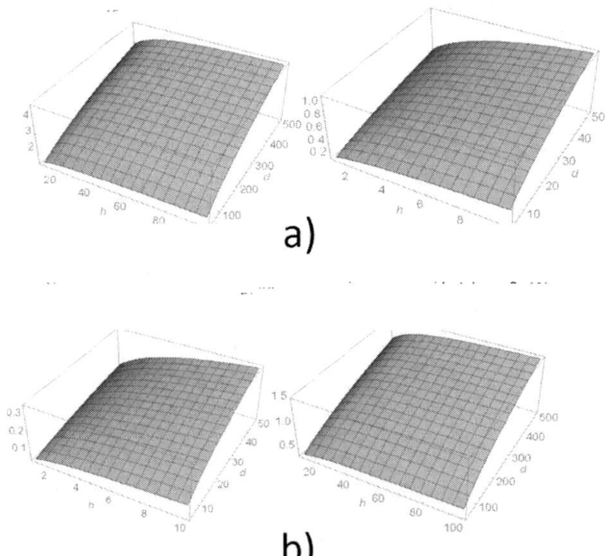

Figure 55. Dependence of the water velocity in the pipeline on its diameter d and height h at different β. a) Water velocity (m/s) in the pipeline on its diameter and height at β=10%; b) Water velocity (m/s) in the pipeline on its diameter and height at β=1%.

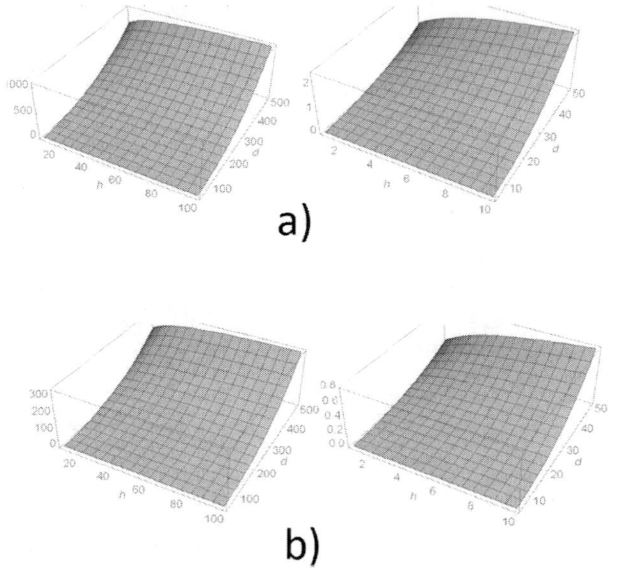

Figure 56. Dependence of the water consumption (m^3/h) on the pipeline diameter d and height h for different β. *a)* Water consumption (m^3/h) on the pipeline diameter and height for β=10%; b) Water consumption (m^3/h) on the pipeline diameter and height for β=1%.

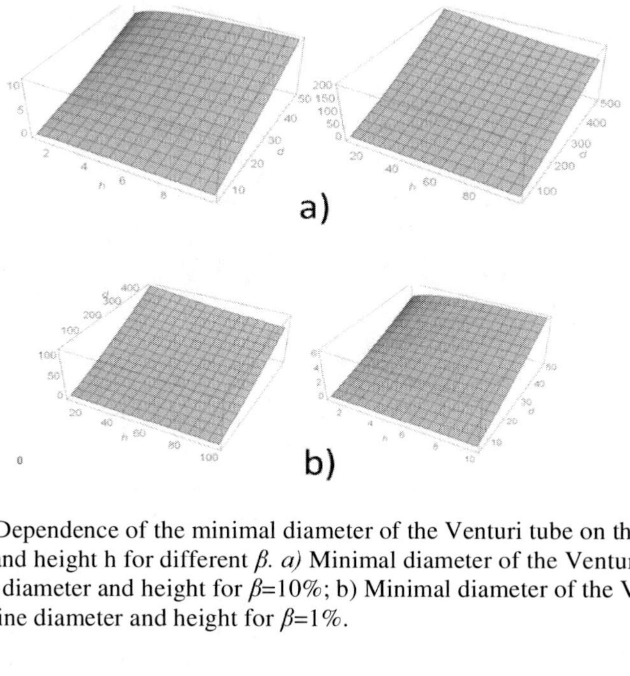

Figure 57. Dependence of the minimal diameter of the Venturi tube on the pipeline diameter d and height h for different β. *a)* Minimal diameter of the Venturi tube on the pipeline diameter and height for β=10%; b) Minimal diameter of the Venturi tube on the pipeline diameter and height for β=1%.

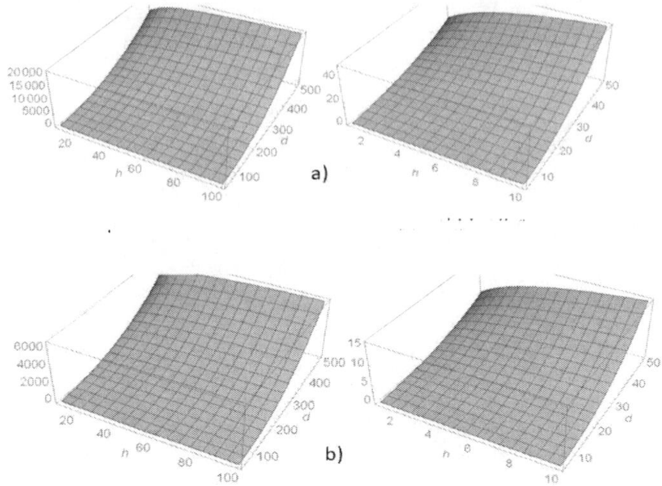

Figure 58. Dependence of the heat power (Kw) on the pipeline diameter d and height d for different β. a) heat power (Kw) on the pipeline diameter d and height h for β=10%; heat power (Kw) on the pipeline diameter d and height h for β=1%.

Centrifugal Devices

The force of gravitation used in the device shown in Figure 53 can be substituted by a centrifugal force. Usually this force is produced by the rotation of the body. For example, if a disk of 1 m radius rotates with an angular velocity of 50 revolutions per second, then the rim of the disk experiences a centrifugal acceleration of about 10^5 m/s^2. In this case, the force acting on any body located on the rim of the disk turns out to be 10,000 times greater than the force acting on this body in the Earth's gravity field.

Thus, the use of centrifugal forces can significantly reduce the length of the pipeline and, accordingly, reduce energy losses arising at a movement of water through the pipeline. Of course, in a short analysis, we cannot consider in detail all the features of such devices, but the main points can be discussed in order to present the principle of operation of such devices.

One of possible scheme of such devices is shown in Figure 59. Here, the Venturi tube 1 rotates around axis 2 in chambers 3 and 4. The inlet fixed pipeline 6 approaches chamber 3 from the left through a centrifugal pump 5. From the left, a curved pipeline 6 enters chamber 3. In chamber 3, the water is transmitted from motionless pipeline 6 to the rotating pipeline 7. Passing through the Venturi tube 1, the water enters through the rotating pipeline 8 into the chamber 4. On the other side of the chamber 4, the water enters through the outlet pipeline 9 again into the water container 10. Motor 11 is used for rotation of the Venturi tube 1 around axis 2.

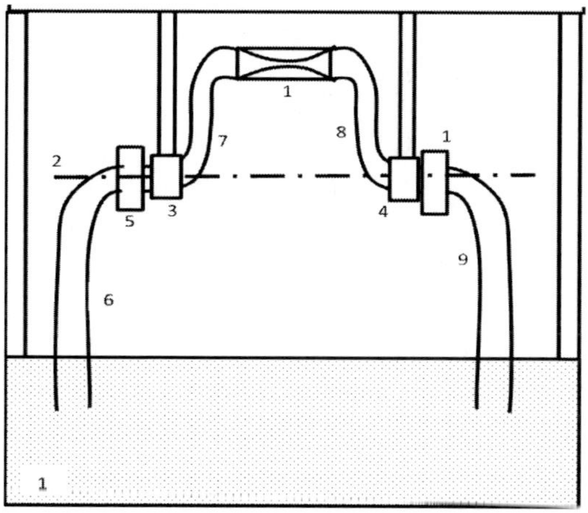

Figure 59. Setup based on centrifugal acceleration, analog setup in Figure 53a based on gravitation.

As is seen, rotating tubes 7, 8 and the Venturi tube 1 in Figure 59 are analogues of pipelines 3, 2 and the Venturi tube 3 in Figure 53. But an acceleration of the Venturi tube in Figure 59 can be by thousands times greater than that in Figure 53. Firstly, this allows to significantly increase in the pressure difference in tubes 7 and 8, that is, to increase the positive pressure that provides circulation of water in the pipeline. Secondly, it allows to significantly reduce the length of the all tubes and, accordingly, reduce the negative pressure associated with inevitable losses in the tubes and in the Venturi tube.

Since the temperature of the water at the outlet of the device is higher than at the inlet, the water container 10 is used as a refrigerator.

This device has the disadvantage that the velocity of the water flowing out of the outlet is different from zero. As a result, the kinetic energy associated with this water is lost.

The question arises, can this disadvantage be eliminated? The answer to this question can be positive if the condition is met that the water at the outlet has zero velocity in a motionless coordinate system. This condition is satisfied in a device similar to the Segner wheel.

Segner Wheel

The Segner wheel is a device for spraying water flowing through a fixed pipe in all directions. Typically, such an installation is used for irrigation. Let us first consider a scheme explaining the principle of operation of such a device.

The device is shown in Figure 60 and consists of a hollow closed cylinder 1 with a round hole 2 at its bottom. Tube 3 is inserted into the round hole 2, the lower end of which is immersed in a container 4 with water. Four identical tubes 5 bent at 90 degrees are inserted in cylinder 1. The container can rotate around a vertical axis passing through the center of the hole 2. Shaft 6 is used to rotate the container 1 and tubes 5 around vertical axis with the help of a motor.

Tubes 5 provides the excess energy. There are two different variants. Firstly, the Venturi tube is used with the possibility of drawing atmospheric air through the holes in the side wall with the formation of air bubbles in the flowing water. The Ventury tube is inserted into the tangential section of the tube 5.

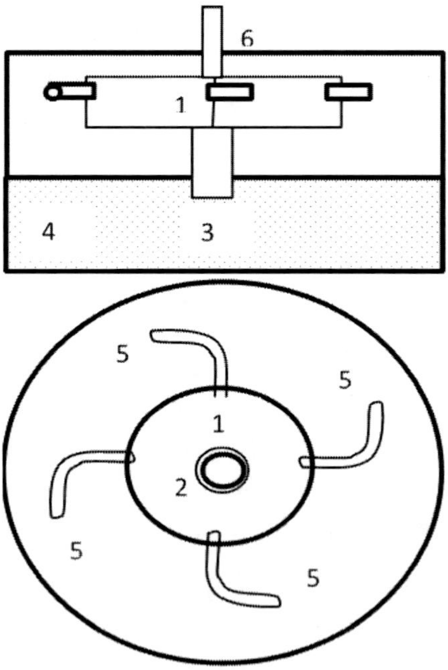

Figure 60. Setup based on centrifugal acceleration and reactive force.

Secondly, there is no Venturi tube, but at the bend of the tube 5 where the pressure of the flowing water is close to zero, holes are made through which atmospheric air is drawn in.

In both cases, when passing through such a section, the density of water decreases due to the Rosch effect, which leads to an increase in the momentum flow of water at the outlet of this section as compared with that of the inlet one.

The operation of the device is as follows. Container 1 is driven clockwise by an external electric motor through shaft 6. Water from container 4 is sucked into container 1 and is thrown out through tubes 5. The water flows down into the container 4 under the action of atmospheric air pressure. The suction of water is explained by the fact that the air pressure at the outlet of tubes 5 is less than atmospheric air pressure, since the speed of the external air relative to the walls of the tube is greater than the speed of the air inside the tubes. According to Bernoulli's law, an increase in the velocity of the outside air is accompanied by a decrease of the pressure in the area where this phenomenon takes place.

Before analyzing the physical principles on which the operation of this device is based, it is necessary to mention several unexpected phenomena arising from the law of conservation of energy. Let us first consider the ideal case, when there are no losses in the system, container 1 rotates at a constant angular velocity, and water flows through tubes 5 at a speed v. Water flows out at the same level at which it flows in. The question is - what should be the speed v? The speed must be such that the kinetic energy of the water flowing out of tubes 5 is equal to zero in a motionless coordinate system. Otherwise, the flowing water will carry away kinetic energy with it, and the energy stored in the system will gradually decrease, which will lead to a slowdown in speed. Thus, the following requirements must be fulfilled. First, the water must flow out in an exactly tangential direction. Second, the speed of the water must be equal to the linear speed of the moving outlet from which the water flows.

These requirements do not depend on the thickness of the tubes, nor on their length, nor on the dimensions of the container 1, nor on the speed of its rotation. Moreover, the thickness of tubes can be variable. What force makes the container rotate? The first set of forces is produced by water when a direction of its movement is changed by 90 degrees. In this case, the radial component of the velocity disappears, but a tangential one appears. These changes are produced by forces acting from the side of the tube walls, which provide a change in these components. Force $F1$, responsible for the disappearance of the radial component of the water velocity, is directed towards the center of rotation. The force $F2$ responsible for the appearance of the tangential component of the water velocity is directed along the water velocity. In accordance with Newton's third law, forces -$F1$ and -$F2$ act on the walls of the tube on the side of the water. Force -$F1$ tries to stretch the tube and is directed from the center. Force -$F2$, which we will refer to as the *pushing force,* is directed against the movement of water, that is, it creates a torque that increases the speed of rotation. It is easy to see that these forces are equal in magnitude, that is, the force that ensures the rotation of the installation is equal to the force that stretches the tube.

Then another question arises. The *pushing force* tries to increase the speed of rotation. But the rotation speed is constant. What force prevents this? When water moves along the radius from the center to the periphery, the tangential velocity of water increases, that is, the water moves with acceleration in the tangential direction. This acceleration is created by a force acting from the side of the tube walls in the tangential direction, and which is called the Cariolis force. This force, which we will refer to as the braking force, is proportional to the velocity and density of the water and is distributed over the radial section of the tube. This force creates a distributed torque along this section. The total torque is equal to a sum of the torque produced by the *pushing force* applied to the periphery of the device and the torque produced by the *braking forces.* If these torques are the same, then the device will rotate uniformly. This should be true at any rotational speed.

The same takes place if the outlet of the tube is closed. In this the water velocity inside the tube is equal to zero and, therefore, trust and braking s forces are equal to zero. The water pressure increases proportional to the square of the linear tangential velocity. The same statement can be expressed in a different way. The water pressure increases proportional to the square of the radius or/and frequency of rotation.

The next question is this. How will this idealized situation change if the Roch effect is taken into account? As for the braking force, it is proportional to a ratio of the speed of the water and its density. Therefore, the torque of the braking force will remain unchanged. As for the pushing force, it is proportional to a product of the water consumption and its speed. The water consumption in this case remains unchanged, but the speed of the water increases. The pushing force increases and produces the greatest torque if a change in the water speed occurs at the largest radius. In other words, the 90-degree change in in the water speed from radial to tangential direction should be as fast as possible, which means that the bend in the tube should be as sharp as possible.

The next question is much more difficult. How will the pressure be distributed inside the tube if the outlet of the tube is open and the water pressure on the outler is closed to a normal atmosphere pressure of 1 bar?

The answer may seem surprising. The pressure in the pipe will decrease as the point at which the pressure is measured approaches the center of rotation. Let's take an example. If the kettlebell rotates on the rope, then the rope will stretch. The rope will not allow the kettlebell to move in a straight line. Now consider a small volume of water in the tube near its periphery. This volume of water rotates in a circle. Where is the rope that makes this volume of water move in a circle? There is no such rope. But if the water pressure on the surface of this volume, located closer to the center, is less than the pressure on the surface of this volume, located farther from the center, then such a pressure difference leads to a force applied to this volume and is directed to the center. This force replaces the tension in the rope. Thus, the pressure in such a tube decreases with decreasing radius of rotation.

As you approach the center, the water pressure gradually decreases and it can become equal to zero at some point. With further advance to the center, the pressure could become negative. However, there are no negative pressures in nature. In this case, the flow breaks and the laminar flow of water ceases. For now, let's leave this situation aside and consider a situation in which zero pressure was directly at the inlet to the tube, that is, inside container 1. On the other hand, the water pressure is different from zero near walls of the container. How can this be? The answer is the following. The pressure gradually changes in the container near the tube from a large pressure in the container to a small pressure in the tube. Such a change in pressure is necessary in order to increase the radial velocity of water from zero velocity in the container to a sufficiently high velocity in the tube.

Thus, we can create a water pressure close to zero in some section of the tube. The Rosch effect may occur in this area. Moreover, if holes are drilled in the side walls of the tube in this section, then the air will be injected through these holes at a pressure of about 1 bar. In this case, the section under consideration works like the Venturi tube, providing a decrease in the density of water at the outlet of the section. As already noted, this leads to an increase in the momentum flow of water and its kinetic energy.

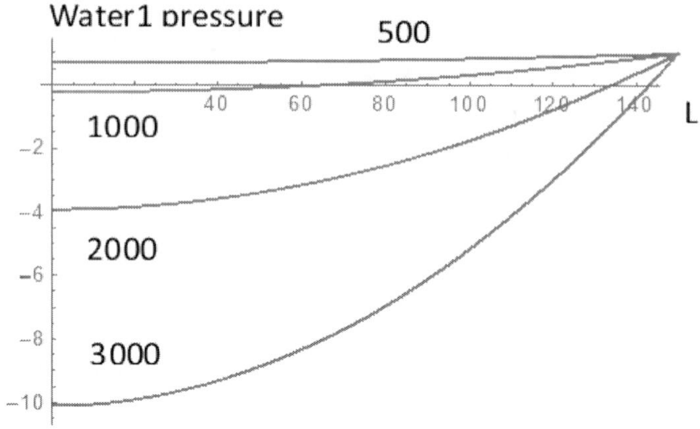

Figure 61. Dependence of the water pressure in the tube on the distance L from the center rotation for various rotations per minute (500, 1000, 2000 and 3000 RPM) for tube of 150 mm length.

Let us consider reasons that violate the balance between the torques produced by the braking and the pulling forces in such a way that the pulling torque surpasses the braking one. This condition will be met if the zero water pressure is implemented as close as possible to the periphery. This condition can be provided by corresponding rotation speed of the container.

That place on the radial section of the tube 5 where the water pressure is equal to zero shifts along the radius with a change in the rotational speed. As an example, Figure 61 shows the location of this place at various rotation frequencies for section of 15 centimeters long. As you can see, at a speed of 500 RPM, such a place does not yet exist. At a rotation frequency of 1000 RPM this place is shifted from the center of rotation by 6 centimeters, with a rotation speed of 2000 this place is shifted by 130 centimeters, at a rotation speed of 3000 RPM this place is at a distance from the end of the section by less than 1 centimeter.

If the place with zero water pressure if shifted to the periphery, the main part of the braking torque will be created in the old way, and the main part of the torque produced by the pulling force, will be created in a new way. As a result, the torque produced by the ejected water jets is greater than the braking torque. The device starts to work independently without the help of a motor. As usual in such cases, the motor can be switched to the electricity generation mode.

The next question that needs to be solved is how does the dimensions of the setup affect its parameters? When the scale of the installation is reduced by a factor of N, all diameters are reduced by N times. In this case, the frequency of rotation must be increased by the same factor of N in order for the operating conditions are not changed. In this case, the speed and pressure of the water on the peripheral are not changed. Losses in the tubes, which are proportional to a ratio of length and diameter are not changed also. Mechanical loads on moving parts created by centrifugal forces are not changed also. However, the power of the setup decreases in proportion to the square of the diameter, that is, by N^2 times. At the same time, the setup volume decreased by N^3 times.

The powers P of the trust and braking forces are identical and is proportional to the square a the tube diameter d and the cube of the velocity v on the outlet of the tubein an ideal case when these powers are identical. For example, if d= 20 мм, and v=20 m/s then P=2.5 kW. However, in the ideal case, these powers compensate each other. If the coefficient $\beta = 0.2$, then the Rosch effect produces a useful power of 500 W. If the losses are 10%, then 250 W remains. The water consumption is 6.3 kg/s.

With this information in mind, one can estimate the water pressure in the container and how much force is applied to the tube. This information is necessary to evaluate the required strength of the setup. Assuming that the water in the container rotates with the same angular velocity as the container, we find that the pressure on the walls of the container is proportional to the square of the radius and the square of the frequency of rotation.

Since the linear speed does not depend on the size of the setup, the dependence of various parameters on speed is valid for installations of various sizes. As an example, Figure 62 shows the dependence of $\Delta R/R$ on the linear velocity v of the periphery of the tube, where ΔR is equal to a difference between the distance from the center of rotation to the maximum radius of the tube and the distance from the center to the place, where the water pressure is zero. As is seen, if the air intake holes are located at distance 0.8 R from the center then ΔR=0.2. Then in accordance with Figure 62 the linear speed v should be about 25 m/s. This velocity can be achieved at various R and angular frequency of rotation. For example, if R=0.1 m, then the length of the circumference is 0. 628 m and the rotation frequency should be about 40 revolutions per second or 2400 revolutions per minute.

It is easy to see that by simply changing the rotational speed it is possible to obtain a situation in which zero pressure in the tube will coincide with the place where the air intake holes are located.

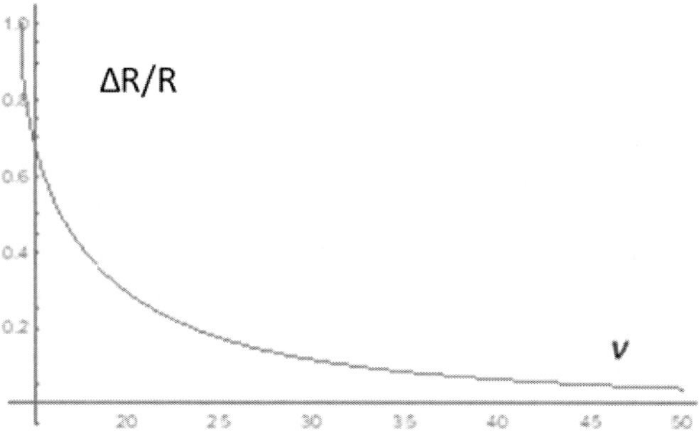

Figure 62. Dependence the distance ΔR along radius on the water velocity in the tube. The distance is measured from maximal radius till the place where the water pressure is equal to zero.

Previously it was shown that the efficiency of the fuel-free device is proportional to the speed of the water v. An increase in speed can be achieved by increasing the radius or/and the speed of rotation. In both cases, the requirements for manufacturing accuracy increase. Therefore, in the manufacture of a particular device, it is necessary to first determine what maximum speed can be achieved with the adopted manufacturing technology. The pressure produced by rotating water within the container is proportional to the square of the linear velocity of its side walls. But the force applied to the tube located at distance R from the center depends also on R. For example, if R=0.5 m, v=50 m/c, and mass of the tube m=10 grams, the force applied to the tube is equal to 2500 Newton or 250 kg of force.

The speed of water at the outlet of the tube is determined by the linear speed of rotation of the rim. In the absence of losses, the thrust torque is equal to the product of the water consumption, the linear speed of the water and the distance between the center of rotation and the outlet of the tube. Assuming that the speed of the water has increased by (1+β) times, where β characterizes an increase in speed due to cavitation, we find that the resulting torque is equal to the difference between the pushing and breaking torques. This difference is proportional to the parameter β.

It remains to consider the requirements for tube 3, through which water is sucked in, and for cylinder 1, in which water acquires a linear speed near the inlet of each of the tubes 5, equal to the linear speed of rotation of these inlets. We have previously shown that changing the density of water should be done as close as possible to the outlet of the tube. The pressure must be equal to zero in this place to exclude a rupture of the water flow in the tube. This means that immediately after this section there should be an entrance to the tube. It follows from this that the linear speed of the inlet differs slightly from that of the outlet. In this case, we have the linear speed of the side walls of the container and the inlets of the tubes differ slightly from the linear speed of the outlets of the tubes.

This means that the water must receive enough a large tangential linear speed inside the container. In this case, the water must spin inside the container due to friction against the side walls of the container. It is also possible that water is sucked in by rotating inlets directly from the tube 3. In any case, the diameter of this tube must be at least 2 times larger than the diameter of the tubes 5 in order to ensure the necessary water consumption. In this case, the water consumption in tube 3 will be equal to that in tubes 5. To increase the water consumption in tube 3, and reduce the pressure drop in this tube, its diameter should be increased.

Another option is also possible, in which water is sucked into each tube 5 through an individual tube 3. In this case, the water is accelerated inside the radial component of the tube 5 due to the action of the Cariolis force that brakes each tube separately.

But as follows from the above analysis, the radial section of the tube must be of a minimum length so that there is no break in the water flow. There is a way out of this contradiction. The radial section of the tube can be made of large diameter. As already noted, this will not affect the total torque created by the Cariolis forces in any way. At the same time, this will favorably affect the losses, since with an increase in the diameter of the tube at the same length, the losses in the tube decrease. What will be the pressure at the junction between thick and thin tubes? It follows from general considerations that if the diameter of a thin tube is sufficiently small, then the pressure in a thick tube will differ little from the pressure created in a thick tube when it is clogged.

Thus, we have a thin tube, at the ends of which there is a large pressure difference. Such a situation is shown in Figure 63a where radial thick tube 2 is connected with curve thin tube 3. The same situation is often considered in hydrodynamics when water flows out of a tank 2 through a long thing tube 3 as is shown in Figure 63b. The water velocity on the outlet of the thin tube is determined by a sum of the pressure produced by the water in the tank and the pressure produced by a colomn of water in the thin tube. The water pressure at the junction of tank 2 and tube 3 in Figure 63b is small. Moreover, if the length of the tube 3 exceeds 10 meters, then a flow break occurs at the junction.

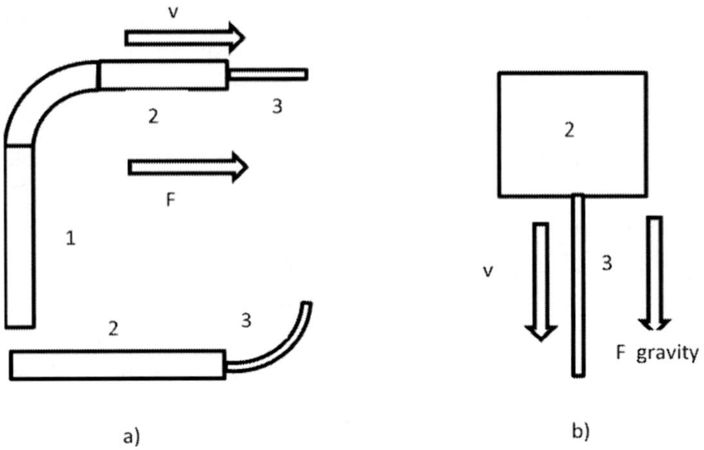

Figure 63. Obtaining high inlet pressure in a thin tube, which is supposed to have low water pressure to suck in outside air. a) diagrams of the installation (view directly and from above), where the pressure is produced by centrifugal force, the direction of which is shown by arrow; b) scheme of the known installation, where the pressure in the thin tube is produced by the weight of water in the gravitational field of the Earth.

A variant is possible in which water enters the inlet of the thick tube 5 under a certain pressure, which ensures the initial filling of the whole pipeline from its inlet to its outlet. This is ensured by bending the lower ends of tubes 5 in such a way that the perpendicular to the inlet plne is moving near the center of rotation and the tangential component of its velocity is different from zero. After the entire pipeline is filled with water, water is sucked into the tube by the low pressure that occurs due to the fact that water pours out of the other end.

As is seen, the principle of operation and the physical phenomena responsible for the operation of this device are exactly the same as in the device discussed above in Figure 53 that is completely submerged in water and located in the Earth's gravity field. The only difference is that in the case under consideration, the force of gravity is substituted by the centrifugal force arising during the rotation of the container and is proportional to the square of the rotation speed. In other words, the *gravitation effect* is substituted by the *centrifugal effect*. In both cases, there are the *momentum effect* and *excess energy effect*. The Rosch effect is responsible for arising these effects.

Of course, the real device must be balanced and consist of many such tubes, evenly distributed in the circumference. Further consideration of these issues applies to engineers and inventors.

Ship Mover

A very curious situation arises if the device in Figure 53b is placed horizontally, as shown in Figure 64.

In this case the *gravitational effect* disappears. But the *momentum and energy effects* leave. As a result, the device will operate but its efficiency can be decreased because of absence of one positive effect connected with a disappearance of the *gravitation effect* but its efficiency can be increased because the negative effect connected with losses of pressure and energy in vertical tubes is absent.

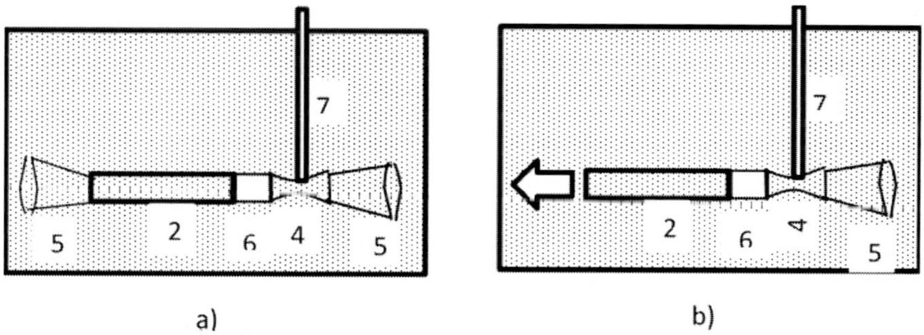

Figure 64. Behavior of the horizontal Venturi tube in water. a) water moves inside the tube; b) the tube moves in motionless water.

Let us first consider the case when the Venturi tube is motionless, as shown in Figure 64a, and a water flow through the tube provides The Rosch effectin its narrow part. The *momentum effect* and *energy effect* continue to work. Since the momentum of the water at the outlet is greater than at the inlet, a force equal to a change in these momentums acts on the flowing water. If this force is greater than the frictional force that occurs when water flows through the tube, we will get a unique device that provides a positive momentum to the water flowing through it. In other words, the Venturi tube creates a positive pressure on the water flowing through it. If this pressure provides the flow velocity necessary for arising the Rosch effect, we will get a unique device that pumps water through itself indefinitely without additional energy sources.

Let us consider the situation when the Venturi tube is immersed in motionless water as is shown in Figure 64b and is accelerated in the direction shown by arrow to the speed v at which the Rosch effect takes place. In this case, there is no need for a funnel 5, since the tube penetrates the still water and the water inside the tube remains motionless relatively the water surrounding the tube.

In the coordinate system connected with the tube, we obtain the same situation as in Figure 64a. A difference is that the tube is not fixed and can move. Since the tube increases the momentum of the water flowing through it, there is the force applied to the water from side of the tube. But, in accordance with the third Newton law for every force applied to some object, there is another force applied to another object. Another force is equal to the first one in magnitude and directed in the opposite direction. In the case under consideration, for the force acting on the water and increasing its momentum, there is another force applied to another object and directed against water flow. The object to which this force is applied is the Venturi tube. If this force is greater than the force of friction between water and the tube, we have the following wonderful situation.

If we immerse the Venturi tube in water in a horizontal direction and accelerate it to such a speed that Rosch effect occurs in its narrow cross section, leading to a decrease in the density of water, then after that the tube will move indefinitely, while extracting energy from the water in which it moves due to the release of the excess energy due to the Rosch effect. We will get a unique device that can move across the seas and oceans for an arbitrarily long time, without requiring any fuel for this, but taking energy for its movement directly from the water in which it floats. With this perspective in mind, this problem should be analyzed more carefully.

Such an analysis cannot be carries out without the use of mathematics and formulas. Since we promised to use no single formula in the book, we will limit ourselves to general consideration only. On the one hand, it is known that the power of losses in a pipe, in general, and in the Venturi tube, in particular, is equal to a product of velocity, water consumption and dimensionless parameter R, that determines total losses. On the other hand, the positive power produced by the Rosch effect is equal a product of the same parameters and dimensionless parameter W. Positive pushing force takes place under the condition $W>R$. Parameter R for the Venturi tube depends on a ratio between its great and small diameters that we designate by α. Dependence $R(\alpha)$ is shown in Figure 65.

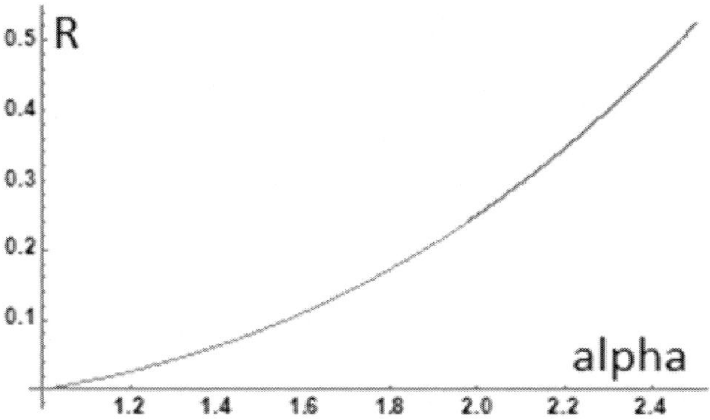

Figure 65. Dependence of parameter *R* on parameter α.

In its turn, a required value of parameter α for which the Rosch takes place depends on the velocity of the Venturi tube relatively a motionless water. This dependence is presented in Figure 66. As is seen in Figure 66, the smaller parameter alpha the smaller losses in the Venturi tube. In the same time the smaller parameter alpha the greater velocity of water on the inlet of the Venturi tube is required.

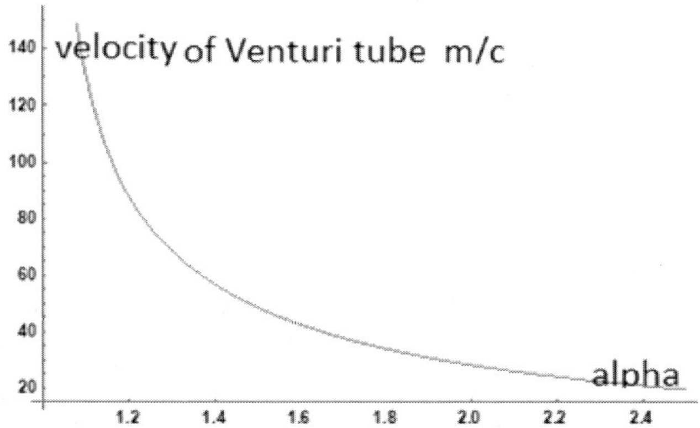

Figure 66. Dependence on parameter alpha of the required velocity of the inlet of the Venturi tube to provide cavitation in the narrow region of the Venturi tube.

As is seen in Figure 66, the greater parameter α the smaller velocity is required to provide the Rosch effect in the Venturi tube. But in accordance with Figure 65 resistance of the Venturi tube increases with an increasing parameter alpha. This means that requirements for the Rosch effect increases with decreasing velocity of the tube. For example, if α=1.8, then in accordance with Figure 66, the velocity of the Venturi tube should be 40 m/s or 144 km per hour. As is seen from Figure 65, resistance *R* in this case is equal 0.2. This means that the volume of water on outlet of the Venturi tube should be greater by 20% than the volume of water on its inlet due to the Rosch effect. As is seen, the smaller requirements to the Rosch effect the greater velocity is required.

An increase of the energy due to cavitation have been achieved in vortex heat generators at *R*=0.5. In this case in accordance with Figure 65 alpha=2.4 and in accordance with Figure 66, velocity of the Venturi tube is equal 20 m/s or 72 km/hour. Since there is currently no theory for obtaining energy by means of the Rosch effect, a final answer can be obtained by means of experimental investigations only.

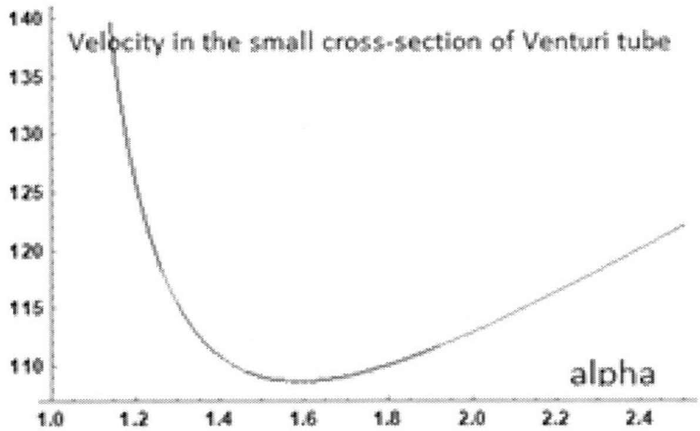

Figure 67. Dependence on parameter alpha on the velocity in the small cross section of the Venturi tube that is necessary to provide cavitation.

It turns out that the velocity of water in the small cross-section of the Venturi tube that is necessary to provide cavitation has minimum as is shown in Figure 67.

An interesting design is shown in Figure 68, where a propeller 2 is inserted into the Venturi tube 1 with the possibility of rotation from an external energy source. Tube 3 is designed for the possible supply of air bubbles. In this case, the possible occurrence of cavitation on the propeller blades favorably affects the increase in its thrust. Once the propeller has propelled the ship to the required speed, the propeller can be turned off. The ship will be pushed by the force resulting from the increased momentum of the water due to the Rosch effect in the narrow section of the Venturi tube. At best, this force may be sufficient to maintain the speed of the ship. In this case, the propeller can be rotated by the flow of water flowing through the Venturi tube and can be used as an electricity generator.

If this force is not sufficient to maintain the ship's speed, the propeller can add an additional thrust. Thus, in this case, the force generated due to the Rosch effect helps the propeller and saves fuel.

As another alternative, a variant can be considered when additional thrust is created by pumping air bubbles through tube 3. Again, the resulting decrease in the density of water with air bubbles contributes to an increase in the kinetic energy of water flowing through the tube, which leads to an increase in its momentum and arising the force that ensures the movement of the Venturi tube.

Our final conclusion is the following. The greater an increase the energy due to the Rosch effect the smaller the velocity of the Venturi tube is required to move without external sources of the energy.

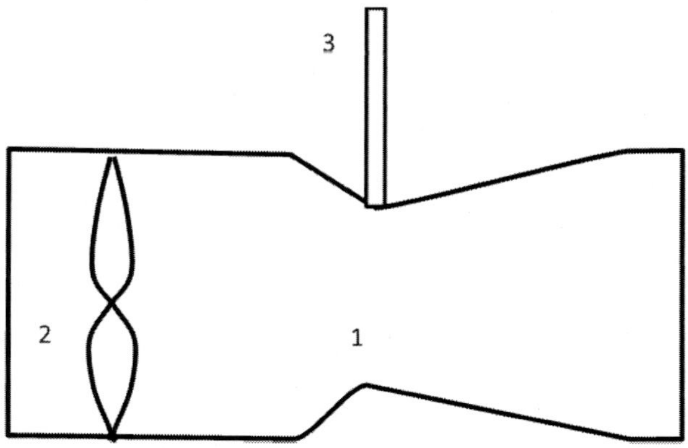

Figure 68. Venturi tube with propeller of sea ship.

Conclusion/Remarks

The purpose of this Chapter was to analyze the possible applications of the circulating light. It is known that currently there are problems with lack of energy in the world. Therefore, attempts to consider the possible consequences of new phenomena discovered in recent times are fully justified. The circulating light has unique parameters regarding gas pressure, energy density and lifetime. Perhaps these new phenomena can be used to solve the energy problems. We have tried to consider the prospects for the use of these new phenomena.

From the presented analysis, we can conclude that the circulating light can contribute to the development of devices in which excess energy appears, extracted from the energy of nuclei during their transformation under the conditions that exist in anomalous objects in the world of the circulating light.

Conclusion about Circulating Light

We can conclude that there are perfectly new objects and phenomena in the new world of the circulating light, about the existence of which nobody earlier knew and suspected even.

At the same time, the inhabitants of this world have long been known, studied and described, but the physical nature of these inhabitants was unclear. That is why they are called anomalous.

It turned out that the anomalies of these inhabitants can be explained on the assumption that the light propagates along their surface in all possible directions. An explanation of the circulating light nature is difficult because everyone knows the properties of ordinary light, and for thousands of years, there has not been the slightest mention about the circulating light. We tried to convince the reader that, in addition to the familiar world with the light around us, there is a world of circulating light in which this light can be compressed and concentrated to a limited volume like air is concentrated in a car tire or water in a vessel.

Unlike ordinary white light that surrounds us during the day, circulating light has unique properties. The most significant from them is its extremely high energy density. This leads to the fact that, unlike all known objects of our world, an additional type of force acts on the circulating light. These optically induced forces are mainly responsible for the observed anomalies in the behavior of Ball Lightnings, which are the most familiar inhabitants from the world of circulating light.

It is shown that inhabitants of different types, sizes, with different lifetimes and stored energy live in the world of circulating light. For brevity, we assigned a nickname for all the inhabitants of this world "ball light." Its shell can also be very diverse. It can be composed of ordinary compressed air, compressed gas, the reflective index of which is greater than that of the surrounding space, molten glass, which reacts with the oxygen of the environment, resulting in a long-term existence of circulating light.

A ball light can be in the form of a transparent liquid droplet where the intense light is circulating along its surface in all possible directions.

A ball light can be in the form of molten material, wherein the exothermic chemical reaction occurs with the ambient air on its surface. The reaction provides the long luminescence of the ball light for about ten seconds.

A ball light can be in the form of a luminous ball, produces by continuous or pulse gas discharge.

The anomalies of various abnormal glowing objects, obtained by researchers in the laboratory over the past two centuries when they tried to produce artificial Ball Lightning, are explained by the presence of the circulating light in these objects.

Certainly, ball lights obey the total physical laws. At the same time, the circulating light is subject to optically induced forces that were known before but are not taken into account because of their weakness. The energy density of the circulating light increases by billions of times as compared with that of the conventional light propagating linearly. This entails the corresponding increase of the optically induced forces, and they begin to play a decisive role in the behavior of the ball lights.

The most famous representative among ball lights is Ball Lightning. Few have seen it, but there is ample eyewitness evidence as everyone is keen to share such an amazing observation. In the first chapter, we showed that all the puzzles and anomalies of Ball Lightning are explained by the presence of the circulating light in it. Moreover, we predicted the behavior of Ball Lightning in situations in which it was not observed. We presented the proof of similarity between behaviors of natural Ball Lightning and the circulating light. We consider this fact as an experimental confirmation of our theory of the circulation light.

Other representatives from the world of circulating light are luminous balls with anomalous properties. Unlike natural Ball Lightning, the behavior of such balls can be seen in the video on the Internet. Moreover, if you wish, you can produce such balls yourself. Their anomaly is that they bounce off a simple sheet of paper more than 10 times, in contrast to ordinary balls, which cannot bounce even once.

We have shown the presence of the circulating light in these balls by using different obstacles that the balls bounce off. The physical properties of these obstacles were chosen so that the interactions of the circulating light with such obstacles were different. All theoretical expectations were completely justified.

Another anomaly of these balls is their slower movement between two adjacent bounces. The time between these bounces turns out to be about two times longer than the estimated time for the movement of conventional bodies. It has been shown that the circulating light is also responsible for this anomaly.

Other objects from the world of circulating light have been obtained when trying to produce Ball Lightning in the laboratory. The presence of the circulating light is manifested in certain anomalies of these objects. Properties of these objects were known and can be analyzed in detail. A simple study of this information enables us to discover several new phenomena and to expand knowledge about the world where we live.

The study of the circulating light was not limited to the topics presented in the book. Theoretical issues were also considered in the articles presented in the Reference section. In our opinion, the main achievement is the proof that the necessary conditions under which the circulating light can exist and be stable are derived and and physical parameters of the air pressure and energy density of the circulating light required by these conditions are determined.

At the same time, two theories have been re-developed. One of them is related to the analysis of the phenomenon, which is currently called electrostriction pressure. It is shown that even the name itself incorrectly reflects the essence of this phenomenon. More correctly, this phenomenon should be called electromagnetic compression of a gas in a field of intense electromagnetic wave. The second theory is related to the analysis of optically induced forces. On the basis of more than a dozen thought experiments, where no assumptions about the physical nature of optically induced forces were used, it has been shown that the momentum of light in an optical medium increases by a factor of n, where n is the refractive index of the optical medium. The causes of this phenomenon are given. Several dozen articles have been published on this subject, each of which contains more than a dozen different formulas. However, these issues should be considered in another book, aimed at a different circle of readers.

The nature of Ball Lightning in the form of the circulating light had been disclosed for about 20 years. However, it cannot be said that the number of supporters of this theory is growing in the world. The famous Copernicus theory that the Earth revolves around the Sun waited its recognition above the century. The new theory is usually not accepted by contemporaries. New generations, comparing existing theories, give preference to the most correct from their point of view.

In the same time, the theory gives positive results at present time. We have already shown that attempts to carry out a thermonuclear reaction based on the phenomenon of sonoluminescence are futile since the visible shift of the radiation spectrum to the short-wavelength region is associated with the compression of the circulating light, and not with radiation of the body heated to a very high temperature. In addition, we have shown that the circulating light must be taken into account in high-risk situations such as flying airliners and nuclear power plants.

Besides, without waiting for universal acceptance, the world of the circulating light allows a deeper understanding of known phenomena and a broader understanding of the laws of the world around us. We have just begun to study the phenomena in the world of circulating light. These phenomena require further study. For example, the world of nuclear physics was discovered at the end of the 19th century. Only half a century later, these phenomena have found practical application.

The knowledge that circulating light exists in nature allowed us to take a fresh look at already known events and phenomena and evaluate the possible participation of circulating light in them. One such event is the accident at the Chernobyl nuclear power plant. Many facts and events are known immediately before the accident and the consequences of the accident, which are not related to each other and do not at all relate to the official explanation of the causes of the accident. However, if we assume that the cause of the accident was Ball Lightning, which occurred a minute before the accident at the beginning of the planned electrical tests, then these facts and events irrefutably indicate that Ball Lightning penetrated the mine with fuel elements and brought the nuclear power plant out of the normal mode, which led to to an explosion due to an increase in vapor pressure above the allowable. An analysis of the possible influence of erupted Ball Lightning on the causes of the accident is given in Appendix 1.

Another anomalous phenomenon, which can be explained taking into account the properties of the circulating light, is excessive heat generation in vortex cavitation generators. An attempt has been made to explain this phenomenon by taking into account the properties of circulating light to compress gases to extremely high pressures. In this case, it is assumed that the circulating light arises in cavitation bubbles. Circumstances are considered under which high pressure can contribute to the occurrence of nuclear substitution reactions, in which the transformation of two identical nuclei occurs. The transformation consists of transferring protons and neutrons of one nucleus into another. Examples of such reactions are given, which are accompanied by a partial conversion of nuclear energy into thermal energy.

It would be unnatural to abandon attempts to analyze the possible applications of new phenomena for practical applications. First of all, attention was paid to anomalous phenomena associated with the appearance of excess energy. At the same time, such a fact was noted that the mention about the excess energy was often accompanied by the mention of luminous anomalous objects from the world of circulating light. Knowing the physical parameters of anomalous objects, it was natural to assume that these parameters are somehow connected with the appearance of excess energy.

This assumption is not unfounded. It has been previously shown that circulating light is responsible for the phenomenon of sonoluminescence. It was substantiated that the glow of a cavitation bubble comes from the molecular scattering of light circulating in the shell of the glowing bubble. Similar phenomena can also occur in cavitation bubbles obtained by other means. Based on these assumptions, possible phenomena were analyzed that allow the known Rosch setup to operate without the apparent use of external energy. It is shown that the circulating light that appears in air bubbles when air is pumped into water could take part in the operation of the setup.

Based on these assumptions, possible phenomena were analyzed that allow the known Rosch setup to operate without the apparent use of external energy. It is shown that the circulating light that appears in air bubbles when air is pumped into water could take part in the operation of the setup. This assumption is justified by the fact that when studying an anomalous object from the world of circulating light in the form of a light bubble that occurs during the phenomenon of single-bubble sonoluminescence, it was found that the glow of the bubble comes from molecular scattering of light circulating in the shell of the bubble.

Bearing in mind that the operation of the Rosch setup without visible external energy sources is an objective reality, and assuming that circulating light is a means for obtaining excess energy, it would be unjustified to consider means that significantly increase the efficiency of this installation and can be used in other areas of application. . Several devices are presented that operate without the use of known energy sources. The most impressive from our point of view is a hollow tube that can move endlessly through the seas and oceans, while consuming energy from the water in which it moves. This device looks fantastic. But the world of circulating light is also from the realm of phantasticism. However, both are based on currently accepted laws and phenomena that cannot be denied.

Appendix 1: Ball Lightning as a Reason for the Chernobyl Tragedy

There are more things in heaven and earth, Horatio,
Than are dreamt of in your philosophy
Hamlet, William Shakespeare

Introduction

Bearing in mind the presence of circulating light in nature and its unusual properties, we were able to explain the nature of many phenomena that were previously considered anomalous. Taking into account the circulating light can be explained from a scientific point of view, without resorting to various assumptions, the causes another phenomenon that is known to all inhabitants of the Earth. This refers to the explosion at the Chernobyl nuclear power plant, which led to the pollution of large areas and significantly affected the development of nuclear energy.

In 2006, when it was 20 years since the tragedy, we published an article in which the hypothesis was substantiated that the cause of the accident was the ball lightning, which occurred when switching large currents at the start of the planned tests of conventional electrical equipment. In a short article, it was impossible to present all the properties and features of the ball lightning that contributed to its occurrence and destructive effect. Only links were provided where the mentioned properties are described.

Now, in the conclusion of this book, where the properties of objects from the world of circulating light were described in detail, it is appropriate to cite the main provisions of this article, because the reader is already familiar with the properties of the Ball Lightning.

An analysis of the causes of the Chernobyl tragedy has been the subject of much research. The official version is that the tragedy is the result of personnel errors, exacerbated by a fateful combination of circumstances and some design flaws of the reactor. The fact that such personnel errors and structural flaws were is not in doubt.

But was this the real cause of the accident? It is impossible to prove or disprove it. This is a simple example. There was a traffic accident on the highway, in which the car was in a ditch. Examining the causes of the accident, the traffic police found that the driver had significantly exceeded the speed, and his car was technically faulty: bad brakes, unacceptable play in the steering, etc. All of this could lead to an accident. However, more thorough studies showed that immediately before the accident, a truck was moving in the oncoming lane towards the car, and the driver had to turn into a ditch to avoid a collision.

The question arises. Was there some unknown reason, similar to a truck moving in an oncoming lane?

From the chronology of events, it is known that about a minute before the explosion, safe pre-planned electrical tests of the non-nuclear part of the station began. The tests began, and a minute later, the station was in ruins. Official versions do not establish a causal relationship between these events. Also, there is no such connection in the work of B.I. Gorbachev "Analysis of the causes and realistic scenario of the Chernobyl accident. The main choice is between the two versions" [1].

An attempt to establish such a connection was made by an employee of the Kurchatov Institute, Doctor of Physical and Mathematical Sciences L. I. Urutskoev in the article "Chernobyl may happen again. The mystery of a nuclear disaster is close to a solution" [2]. He believed that "the catastrophe trigger was apparently an experiment that did not portend anything bad when starting cooler pumps in an artificially created emergency mode" [2].

However, the causes of the accident that he describes, associated with the occurrence of magnetic monopoles at the beginning of electrical tests, which penetrated the pipelines into the nuclear reactor and brought it out of normal operation, are currently not supported by either theoretical or experimental data. At the same time, he noted the following two very strange circumstances that he cannot explain.

1. "A great circle of *a regular round shape*" in a sheet of metal in the wall of the shaft of the reactor was burnt by an unknown object
2. The walls of the reactor shaft of the 4th power unit were covered with an oil emulsion. It is known that the core of a nuclear reactor does not require lubrication and the reason for the appearance of oil emulsion on the walls of the mine is a question that no one can answer.

The question about the reason for the appearance of the great circle of a regular round shape belongs to the same series. Since there are no answers, these circumstances were ignored.

In the case at hand, the oddities seem unique. If the hole was of some kind of irregular shape, it would not be as strange as a hole of the correct round shape. If the walls of the mine were covered with soot, it would not be as strange as the walls covered with oil emulsion. If the explosion occurred a few hours after the start of the test, then the connection between these events could be ignored.

If a reasonable explanation can be found linking these inexplicable phenomena into a single whole, then such an explanation can with a high degree of certainty be considered as a true explanation of the cause of the accident.

In this publication, we show that these three circumstances, as well as several other circumstances described by eyewitnesses of the tragedy, could have occurred in a single case when the Ball Lightning appeared while the pumps were turning off. At this moment, the switching of large currents occurs. As was shown in Chapter 2, the same situation occurs during an erosive gas discharge where the transformer oil poured into a high power oil switcher played the role of a hydrocarbon in the erosive gas discharge.

Just as the well-known detective Sherlock Holmes restored the picture of the crime and calculated the criminals from a few minor details, we, from minor details that at first glance are not related to each other and were ignored due to the impossibility of explaining them, gave an explanation in which these details turned out to be interconnected and should have appeared as a result of the occurrence of the Ball Lightning.

An Actual Description of Events before an Explosion

It makes sense to provide some factual material and eyewitness accounts of these tests [1]. Electrical tests began at 01 h. 23 min. 04 sec. A head of the pre-emergency evening shift of the fourth block Y. Treguba testifies:

> "The run-out experiment begins. They disconnect the turbine from the steam, and at that time, they watch how long the run-out will last. And so the command was given .... We did not know how the run-out equipment works, therefore, in the first seconds, I perceived that "there was some kind of bad sound as if the Volga car started to slow down and started to move with a skid. Such a sound: do-do-do-do, turning into a roar. There was a vibration of the building ... But not like an earthquake. If you count up to ten seconds - there was a rumble, the oscillation

> frequency fell. And the power of oscillations grew. Then the blow sounded ... This blow was not very strong. Compared to what was after. Although a strong blow. The control room was shaking. I bounced, and at that time, the second blow followed. Now it was a very strong blow. Stucco fell, the whole building shacked ... the lights went out, then emergency power restored ... Everyone was in shock ..." [1].

Here is how B.V. Rogozhkin describes the course of the accident, who worked on an emergency night as the shift supervisor of the station.

> "All events took place within 10-15 seconds. Some kind of vibration appeared. The hum rumbled rapidly. The power of the reactor first fell, and then began to increase, not giving in to regulation. Then - a few sharp pops and two "hydro shocks." The second more powerful - with the side of the central hall of the reactor. On the block billboard, the lights went out, the slabs of the suspended ceiling fell, all the equipment turned off" [1].

The shift supervisor Orlov recalls that at first, he heard a rumble similar to the sound emanating when one of the parts of the turbogenerator went out of operation, Then, according to him, a fire broke out at the lower elevations - transformer oil was burning. Other employees of the power plant recall the same thing. (We will take into account the burning of transformer oil as the fourth circumstance, which was not taken into account when analyzing the causes)

How could the harmless fact of disconnecting the turbine from the steam affect the nuclear reactor, bearing in mind that the turbine is not in close proximity to the reactor? Having known the properties of the Ball Lightning and the reasons for its appearance, we concluded that there is good reason to believe that the cause of the accident was the Ball Lightning, which occurred when the generator was disconnected from steam during high current switching.

How to Integrate Independent and Unexplained Circumstances into One

Thus, we have the following 4 unrelated circumstances:

1. The station was in ruins within one minute after the start of testing
2. A great circle of a regular round shape in a sheet of metal in the wall of the shaft of the reactor was burnt by an unknown object
3. The walls of the reactor shaft of the 4th power unit were covered with an "oil emulsion."
4. The burning of transformer oil at the lower elevations.

We can take into account the following properties of abnormal glowing objects (AGO) considered in the book:

1. Beneficial effect of the erosive gas discharge on AGO formation.
2. AGO properties to burn holes of regular round shape in a copper foil.
3. Ability of AGO to move in the direction of increasing gas density or gas pressure. As was shown in Chapter 1, AGO moves in the opposite direction to the gas flow in a pipeline.

In this case, we can link together all 4 unrelated circumstances as follows.

AGO was produced due to the erosive gas discharge during high current switching when the turbine was disconnected from the steam.

Of course, this AGO was significantly superior in stored energy to the AGO studied in the laboratory, where the discharge energy does not exceed several hundred joules. The discharge energy during switching currents in a system of power tens of megawatts can be hundreds of thousands of times greater.

It is known that Ball Lightning occurs on submarines when switching electric motors. Ball lightnings are floating inside and burn metal partitions. However, the power of electric motors in submarines is much smaller than that of electric motors in nuclear power plants.

Accordingly, the energy stored in the arising AGO can be comparable with the energy of natural Ball Lightning. As shown the AGO shell tends to draw in gases whose reflective index is greater than the air reflective index. The reflective indices of hydrocarbon vapors are greater than the air reflective index. Therefore, the vapors of the transformer oil are drawn into the AGO shell.

A powerful AGO after its rise begins to move according to the same laws as natural Ball Lightning, namely in the direction of increasing the reflective index of the medium in which it is located. Having penetrated the steam line, the AGO moves in the direction of increasing vapor density. The density of the vapor increases with increasing pressure. In any pipe through which the gas moves, the pressure decreases as the gas moves. The pressure is maximal at the end of the pipe where the gas comes in. These are fuel elements. Having destroyed one of the elements, the AGO entered the mine, where it attracted to the metal wall and burned a hole of the correct round shape in it. Similar holes of regular round shape, burnt in targets made of electrically conductive materials (aluminum, titanium, tungsten) can be seen in the photographs shown in [3].

When AGO disappeared, the vapors of the transformer oil contained in its shell were scattered and settled on the walls of the mine. This explains the fact that the walls of the mine were covered with an oil film. Thus AGO was that trigger, after which the accident occurred. We could finish the description of the causes of the tragedy, but 2 more circumstances were noted by witnesses about the course of the subsequent development of events.

Thus, the appearance of Ball Lightning led to an emergency that could not be foreseen by the operators or the designers. There was a depressurization of the system, providing a constant flow of steam into the turbine. In this case, relaxation oscillations could occur. Such oscillations are observed in nature very often. The creaking of doors, the squealing of brakes, the sounds made by wind musical instruments are also caused by relaxation oscillations.

Many heard the "do do do" sounds when they opened the faucet in the kitchen. At the same time, sounds are heard in neighboring apartments. These sounds occur when the laminar flow of water is interrupted by a bad water valve. The pressure in the water supply is only a few atmospheres. The vapor pressure in the steam line of the nuclear power plant is about 50 atmospheres, and the diameter of the pipes is hundreds of times greater. Therefore, the sounds made by relaxation oscillations arising as a result of a steam line malfunction are much more powerful.

Let us assume that you buy a new table in a store and begin to push it along a street. You know your action is dangerous for the table as it will eventually fall apart, but it will fall apart gradually. First, there will be a high-pitched sound like "r –r –r." Then, as the attachment of the legs to the table decreases, the frequency of the sound decreases, but the amplitude of the vibration increases.

A consistent increase in the amplitude of jumps of the table legs and an decrease in the frequency of jumps in the case when their fastening to the tabletop is gradually decreased in the process of pushing the table along a horizontal surface.

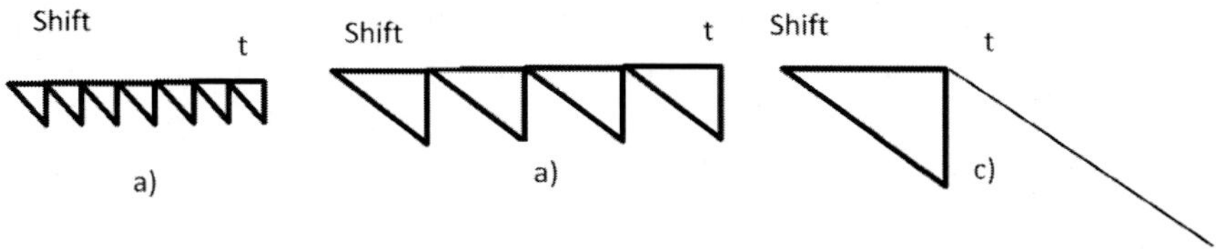

Figure A-1. A consistent increase in the amplitude of the jumps of the table legs and an increase in the frequency of jumps when they are loosened to the table top in the process of pushing the table along a horizontal surface.

This situation is illustrated in Figure A-1 where a change in time of the horizontal offset of the lower end of the table legs relative to its tabletop is shown.

Initially, when the tabletop begins to move in a horizontal direction, the lower ends of the legs remain motionless due to friction against the road surface. Since the legs are firmly attached to the tabletop, with a slight displacement of the tabletop, a force arises that overcomes friction forces and the legs move relative to the tabletop in in its initial position. The displacement of the legs relative to the tabletop becomes equal to zero. After that, the process is repeated. This situation is shown in Figure A-1a). If the fastening of the legs to the tabletop weakened, then the force that overcomes the friction force arises with a greater displacement of the tabletop. This situation is shown in Figure A-1b). With further weakening of the fastening of the legs, their displacement relative to the tabletop becomes even greater. Ultimately, this displacement becomes unacceptably large, at which the legs fall off the tabletop. This situation is shown in Figure A-1c).

You will hear a sound like "do do do." Ultimately, one of the table legs falls off, and the table "explodes," while the state of the table changes noticeably.

Thus, Ball Lightning, which penetrated the steam line, led to a disruption in the normal operation of the steam line. As a result, relaxation oscillations appeared in the system. These oscillations gradually increased the violation. As with the increase in the violation of the attachment of the table legs to the tabletop, the amplitude of these oscillations gradually increased, and the frequency decreased. Eventually, the table was destroyed since its legs and tabletop were separated. Similarly, an increase in the amplitude of relaxation oscillations led to the destruction of the steam line.

Orlov described a similar picture before the explosion. At first, "some kind of bad sound appeared, as if the Volga car began to slow down at full speed and began to move with a skid. Such a sound: doo-doo-doo-doo ... "The frequency of vibrations fell, and their strength grew." These fluctuations led to the failure of the system. "The impact sounded." Thus, the appearance of AGO was the trigger that caused the further course of events.

It can be concluded that disparate exotic unrelated phenomena and events can find a natural explanation, if we assume that at the beginning of the planned tests, Ball Lightning worked, which penetrated the steam pipeline from the fuel element in the mine, destroyed it and disrupted the normal operation of the station. The disturbance gradually increased, which eventually led to the explosion of the system under a pressure of 50 atmospheres. There was no nuclear explosion. A large boiler in the form of a mine room exploded, in which the steam pressure exceeded the permissible limit.

The so-called perpetrators of the accident, who were imprisoned for 10 years, were innocent and were in jail for nothing. We can conclude that the true cause of the tragedy is the lack of knowledge among humankind about natural phenomena that could affect their dangerous buildings.

Conclusion

Even if the primary cause of the tragedy was not Ball Lightning, it is necessary for the future to exclude the possibility of the occurrence of Ball Lightning at nuclear power plants. At the same time, one must not only take great care in switching high currents but also take measures to prevent natural Ball Lightning from entering the nuclear power plant's room. Ventilation is not permissible, in which excessive pressure is created in the building of nuclear power plants. In this case, all Ball Lightnings appearing near the building will tend to penetrate inside just like they penetrate the cabin of a flying airplane, where the pressure is higher than outside.

References

[1] Gorbachev B. I. The causes of the Chernobyl accident: the final choice between the two versions. *Collection "Problems of Chernobyl,"* issue 10, part 1 of Chernobyl 2002.

[2] Chernobyl may recur. The mystery of a nuclear disaster is close to a solution. In : *New Izvestia,* Russian newspaper /2000/09/05.

[3] Avramenko, R. F., V. I. Nikolaeva, L. P. Poskacheva "Energy-intensive plasma formations initiated by an erosive discharge is a laboratory analogue of ball lightning" In Book: *Ball lightning in a laboratory,* Ed. Avramenko R. F. (Moscow, Himiya, 1994).

Final Remarks

There is currently no doubt that Ball Lightning is an optical phenomenon and is a bubble of light that looks like a soap bubble. It also exists for a limited time, disappears instantly and suddenly on an arbitrary section of its trajectory, leaving no traces. The only difference is that the shell of the light bubble consists of a thin layer of compressed air in which light circulates in all possible directions. The circulating light is not blown away by the wind, is not subject to gravity, and moves according to the laws of optics. This explains all the puzzles and anomalies in the behavior of Ball Lightning in the Earth's atmosphere.

Ball Lightning is only one of the representatives of the unknown and unexplored world of circulating light that exists in nature. Accounting for the existence of this world makes it possible to explain some of the known anomalous and mysterious phenomena and events, as well as to analyze the prospects for using the features of this world for practical application.

References

[1] Singer, S. *The Nature of ball lightning.* Plenum Press: NY, 1971.

[2] Stakhanov I. P. *The physical nature of ball lightning.* (Atomizdat, Moscow 1979 CEGB trans CE 8244).

[3] Barry J. D., *Ball lightning and Bead Lightning.* Plenum Press: NY, 1980.

[4] Steinhoff M. *Ball lightning – An Unsolved Problem in Atmospheric Physics,* Plenum PNY, 1999.

[5] Sagan P. *Ball lightning: Paradox of physics.* 2004 Lincoln, NJ.

[6] Grigor'ev A. I., Grigor'eva L. D., Shiryaeva S. O. Ball lightning penetration into closed rooms: 43 eyewitness accounts, *J. Sci. Explor.* 6(3) (1992) 261-279.

[7] Torchigin V. P. (2003). On the nature of ball lightning. *Doklady Physics,* 48(3), 108–111.

[8] Torchigin V. P. Optical Resonators in the Atmosphere. *Laser Physics,* Vol. 13, No. 6, 2003, pp. 1 14.

[9] Torchigin V. P. Optical Resonators in the Atmosphere. *Laser Physics,* Vol. 13, No. 7, 2003, pp. 919

[10] Torchigin V. P. Propagation of Self-Confinement Light Radiation in Inhomogeneous Air. *Physica Scripta.* Vol. 68, no. 6, (2003) 388.

[11] Torchigin V. P. Mechanism of the Appearance of Ball Lightning from Usual Lightning. *Doklady Physics,* Vol. 49, No. 9, 2004, pp. 494.

[12] Torchigin V. P., Torchigin A. V. Manifestation of Optical Quadratic Nonlinearity in Gas Mixtures. *Doklady Physics,* Vol. 49, No. 10, 2004, pp. 553.

[13] Torchigin V., Torchigin A. Role of Ball Lightnings in Low Energy Nuclear Reactions. *Infinite Energy* 54 (2004) 46.

[14] Torchigin V. P., Torchigin A. V. Space solitons in gas mixtures. *Optics Communications* 240 (2004) 449.

[15] Torchigin V. P., Torchigin A. V. Behavior of self-confined spherical layer of light radiation in the air atmosphere. *Physics Letters A* 328 (2004) 189.

[16] Torchigin V. P., Torchigin A. V. Phenomenon of ball lightning and its outgrowths. *Physics Letters A* 337 (2005) 112.

[17] Torchigin V. P., Torchigin A. V. Features of ball lightning stability. *Eur. Phys. J.* D. 32 (2005) 383.

[18] Torchigin V. P. The ball lightning phenomenon and its solution, *Atomic Energy Bulletin 1* (2006) 57-61 (in Russian).

[19] Torchigin V. P. Is it possible to consider ball lightning as a possible hypothesis of the Chernobyl accident. *Bulletin on Atomic Energy* 4 (2006) 89-92 (in Russian).

[20] Torchigin V. P., Torchigin A. V. Self-organization of intense light within erosive gas discharges. *Physics Letters A* 361 (2007) 167.

[21] Torchigin V. P., Torchigin A. V. *Ball Lightning* (from incredible to obvious) (Moscow, URSS, 2008) (in Russian).

[22] Torchigin V. P., Torchigin A. V. On phenomenon of light radiation from miniature balls immersed in water. *Physics Letters A* 374 (2010) 588-591.

[23] Torchigin V. P., A. V. Torchigin Ball lightning as an Optical Incoherent Space Spherical Soliton. In *Handbook of Solitons: Research, Technology and Applications*. Editors S. P. Lang and Salim H. Bedore. Novapublishers (2010) 3-54.

[24] Torchigin V. P., A. V. Torchigin Ball lightning as an Optical Incoherent Space Spherical Soliton. In book *Lightning: Properties, Formation and Types*. Editor Matthew D. Wood Novapublishers (2011) Chapter 6 133-184.

[25] Torchigin V. P., Torchigin A. V. Interrelation between ball lightning and optically induced forces. *Phys. Scr.* 88 (2013) 035402 (6pp).

[26] Torchigin V. P., Torchigin A. V. Nonlinear properties of gaseous optical mediums in a context of ball lightning explanation. *Optik* 127 (2016) 2319.

[27] Torchigin V. P., Torchigin A. V. Stability of the self-confined light. *Optik* 127 (2016) 2298.

[28] Torchigin V. P., Torchigin A. V. How Ball Lightning penetrates in room through small holes and splits. *Optik* 127 (2016) 6155.

[29] Torchigin V. P., Torchigin A. V. Ball Lightning as a self-confined light. *Optik* 127 (2016) 2202.

[30] Torchigin V. P., Torchigin A. V. How the ball lightning enters the room through the windowpanes. *Optik* 127 (2016) 5876.

[31] Torchigin V. P., Torchigin A. V. How ball lightning manages to catch up a flying aircraft and penetrate into its salon. *Optik* 148 (2017) 196.

[32] Torchigin V. P. Explanation of abnormal behavior of ball lightning near the earth Surface. *Optik* 171 (2018) 167.

[33] Torchigin V. P. Explanation of anomalous bouncing luminous droplets of liquid silicon in a framework of the optical model of Ball Lightning. *Optik* 171 (2018) 188.

[34] Torchigin V. P. Explanation of abnormal motion of ball lightning near the earth's surface. *Optics* 171 (2018) 149.

[35] Torchigin V. P. Explanation of abnormal motion of glowing silicon balls in a framework of optical model of ball lightning. *Optik* 176 (2019) 704.

[36] Torchigin V. P. How optical model of ball lightning explains its paradoxical movement upwind. *Optik* 184 (2019) 533-537.

[37] Torchigin V. P. Why ball lightning disappears suddenly traceless. *Optik* 187 (2019) 65-69.

[38] Torchigin V. P. How ball lightning finds out slots and holes to penetrate through them. *Optik* (2019) 163126.

[39] Torchigin V. P. Ball lightning as a bubble of self-confined light. *Optik* 186 (2019) 63-71.

[40] Torchigin V. P. Physics of a ball lightning in a form of a bubble of light. *Optik* 194 188 (2019) 294-301.

[41] Torchigin V. P. Ball lightning as a bubble of light: Existence and stability. *Optik* 193 (2019) 162961.

[42] Torchigin V. P., Torchigin A. V. Simple explanation of physical nature of ball lightning. *Optik* 203 (2020) 164013.

[43] Torchigin V. P. Comparison of the stabilities of a soap bubble and the ball lightning in the form of a bubble of light. *Optik* 219 (2020) 165095.

[44] Torchigin V. P. Physical phenomena responsible for stability and spherical form of ball lightning. *Optik* 219 (2020) 165098.

[45] Torchigin V. P. Physical and optical laws responsible for existence and abnormal behavior of ball lightning. *Optik* 222 (2020) 165374.

[46] Torchigin V. P. Steady states and stability of the circulating light in the air atmosphere. *Optik* 235 (2021) 166635.

[47] Torchigin V. P. Sonoluminescence and circulating light. *Optik* 239 (3) (2021) 166799.

[48] Torchigin V. P. Influence of the circulating light on anomalous properties of glowing objects. *Optik* 242(3) (2021) 167090.

[49] Torchigin V. P. Unknown features of a behavior of the circulating light. *Optik* 241 (2021) 167233.

[50] Torchigin V. P. Ball lightning and the Chernobyl accident. *Optik* 241 (2021) 167224.

[51] Torchigin V. P. Parameters of natural ball lightning derived from its optical model. *Optik* 243 (2021) 167390.

[52] Torchigin V. P. Behavior of the circulating light in a mixture of gases. *Optik* 243 (2021) 167398.

[53] Torchigin V. P. Circulating light and a separation of components of a mixture of gases. *Optik* 244 (2021) 167524.

[54] Torchigin V. P. Circulating light that burns out holes in metal sheets. *Optik* 243 (2021) 167528.

[55] Torchigin V. P. Circulating light at continuous electrical current. *Optik* 244 (2021) 167395.

[56] Torchigin V. P., Torchigin A. V. Comparison of various approaches to the calculation of optically induced forces. *Annals of Physics* 327 (2012) 2288 2300.

[57] Torchigin V. P., Torchigin A. V. Interrelation between striction forces in dielectrics and optically induced forces in transparent media. *Phys. Scr.* 86 (2012) 025402 (8pp).

[58] Torchigin V. P., Torchigin A. V. Interrelation between various types of optically induced forces. *Optics Communications* 301–302 (2013) 147–151.

[59] Torchigin V. P., Torchigin A. V. Compensation of the optically induced Lorentz force in a homogeneous optical medium. *Optik* 124 (2013) 5492– 5495.

[60] Torchigin V. P., Torchigin A. V. Interrelation between ball lightning and optically induced forces. *Phys. Scr.* 88 (2013) 035402 (6pp).

[61] Torchigin V. P., Torchigin A. V. Optically induced force in a curve waveguide. *Eur. Phys. J. Appl. Phys.* (2013) 63: 10501.

[62] Torchigin V. P., Torchigin A. V. Magnitude of the photon momentum in matter *American Journal of Science and Technology* 2014; 1(4): 151-156.

[63] Torchigin V. P., Torchigin A. V. Comment on "Theoretical analysis of the force on the end face of a nanofilament exerted by an outgoing light pulse" *Phys. Rev. A.* 89 no.5, (2014) 057801.

[64] Torchigin V. P., Torchigin A. V. Propagation of a light pulse inside matter in a context of the Abraham–Minkowski dilemma. *Optik* 125 (2014) 2687 2691.

[65] Torchigin V. P., Torchigin A. V. Resolution of the Age-Old Dilemma about a Magnitude of the Momentum of Light in Matter Hindawi Publishing Corporation. *Physics Research International.* Volume 2014, Article ID 126436, 9 pages.

[66] Torchigin V. P., Torchigin A. V. The momentum of an electromagnetic wave inside a dielectric derived from the Snell refraction law. *Annals of Physics* 351 (2014) 444 446.

[67] Torchigin V. P., Torchigin A. V. Pressure Exerted on a Semi-Infinite Lossless Dispersionless Dielectric by a Plane Electromagnetic Wave. *Open journal of modern physics* (2014) ISSN (print): 2372-627X ISSN(Online): 2372-6288.

[68] Torchigin V. P., Torchigin A. V. Comment on "Deducing radiation pressure on a submerged mirror from the Doppler shift. *Phys Rev A* 92 no.1 (2015) 01794.

[69] Torchigin V. P., Torchigin A. V. Kinds of optically induced force derived from laws of conservation of the momentum and energy. *Optik* 126 (19) (2015) 1878.

[70] Torchigin V. P., Torchigin A. V. Radiation pressure on plane dielectric surfaces. *Optik 126* (19) (2015) 1767.

[71] Torchigin V. P., Torchigin A. V. Optical pressure applied to a semi-infinitive dielectric. *Optik* 127 (2016) 6066 6070.

[72] Torchigin V. P., Torchigin A. V. Photon and physical phenomena responsible for its momentum. *Optik* 127 (2016) 6120 6128.

[73] Torchigin V. P., Torchigin A. V. Physical nature and magnitude of optically induced forces derived from laws of mechanics. *Optik* 127 (2016) 5976-5983.

[74] Torchigin V. P., Torchigin A. V. Equation of motion of a photon in an optical medium. *Optik* 127 (2016) 6708.

[75] Torchigin V. P., Torchigin A. V. Photon momentum and optically induced force in matter derived from eikonal equation. *Optik* 127 (2016) 8604.

[76] Torchigin V. P., Torchigin A. V. Radiation pressure in dielectrics. *Optik* 2017 May.

[77] Torchigin V. P., Torchigin A. V. Optical electrostriction pressure and its influence on the momentum of the light pulse in an optical medium. *Optik* 148 (2017) 120 123.

[78] Torchigin V. P. Momentum of a wave and additional forces connected with the momentum. *Optik* 158 (2018) 861.

[79] Torchigin V. P. Comparison of various approaches for calculation of optically induced forces. *Optik* 183 (2019) 346.

[80] Torchigin V. P. Additional arguments in favor of the Minkowski form of the momentum of light in matter. *Optik,* Volume 186, June 2019, Pages 390-394.

[81] Torchigin V. P. Momentum of a wave of any physical nature in a context of the Abraham-Minkowski dilemma. *Optik,* 187 (2019) 148-151.

[82] Torchigin V. P. Another argument in favor of the Minkowski form of the momentum of light in matter. *Optik* 194 (2019) 163127.

[83] Torchigin V. P. Mass of the photon propagating in an optical medium and mass of its electromagnetic and mechanical components. *Optik* 194 (2019) 163125.

[84] Torchigin V. P. Optical electrostriction pressure in liquid, solids and gases. *Optik* 189 (2019) 90-96.

[85] Torchigin V. P. Momentum and mass of a pulse of light wave as a particular case of waves of arbitrary physical nature. *Optik* 202 (2020) 163605.

[86] Torchigin V. P. Momentum of an arbitrary wave derived from the Doppler law. *Optik* 204 (2020) 163518.

[87] Torchigin V. P. Dozen arguments in favor of the Minkowski form of the momentum of light in matter. *Optik* 218 (2020) 164986.

[88] Torchigin V. P. Ratio of the momentums of an arbitrary wave propagating in various media. *Optik* 219 (2020) 164987.

[89] Avramenko R. F. Ed. *Ball lightning in a laboratory*. (Moscow, Himiya, 1994).

[90] Paiva G. S., A. C. Pavao, E. A. de Vasconcelos, O. Mendes jr, and E. F. da Silva jr. 2007. *Phys. Rev. Letters, PRL* 98, 048501 (2007) January 2007.

[91] Grigoriev I. S. *Handbook of Physical Quantities,* Ed. (Energoatomizdat, Moscow 1991; CRC Press, Boca Raton, 1997).

[92] Abrahamson J. and J. Dinnis, 2000. ball lightning caused by oxidation of nanoparticle networks from normal lightning strikes on soil. *Nature*, vol.403, 3, February 2000, pp.519-521.

[93] Bychkov V. L., Bychkov A. V., Timofeev I. B. Experimental modelling of long-lived luminous formations based on polymer organic materials. *Zh. Tekh. Fiz.,* vol. 74 (2004), no. 1, 128-133.

[94] Emelin S. E. et al., 1997. Some objects arising from the interaction of an electric discharge with a metal and polymer. *Technical Physics,* v 67, No. 3, 19–28 (1997).

[95] Klimov A. I., G. I. Mishin, Anomalous wave and gas dynamic properties of long-lived energy-intensive plasmoids. *Letters to the ZhTF,* vol. 19 (1993), no. 13, 19-24.

[96] Klimov A. I., Melnichenko D. M., Sukovatkin N. N., 1994. Long-living energy-intensive excited formations and plasmoids in liquid nitrogen. In the book *"Ball lightning in the laboratory."* Ed. R. F. Avramenko. - M.: Khimiya, 1994, p.72.

[97] Dimitrov S. K., Zhdanov S. K., Kirko D. L. In *Ball lightning in a laboratory;* Avramenko R. F.; Ed.; Himiya: Moscow, 1994 p.79.

[98] Klimov A. I., Gridin A. Ju., Mishin G. I. Long-lived plasmoids obtained other means. In the book *"Ball lightning in the laboratory."* Ed. R. F. Avramenko. - M.: Khimiya, 1994 p.175.

[99] Egorov A. U., Stepanov S. I., 2002. Long-lived plasmoids – analogies of ball lightning arising in moist air. *Journal of Technical Physics* 72, (2002), issue 12, 102–104.

[100] Mesyats G. A. *Ectons in vacuum discharge: discharge, spark, arc;* Nauka: Moscow, 2000.

[101] Mesyats G. A., 1995. *Uspehi Physics;*1995 165 (6), 601–626.

[102] Shoulders K., 2005. Charge Clusters, *Infinite Energy,* 61, 2005.

[103] Tsiklis D. S. *Dense gases* (Moscow, Himiya, 1977) (in Russian).

[104] Ellingsen S. A., Brevik I. (2012). Electrostriction counter-force on fluid microdroplet in short laser pulse. *Optics Letters,* 37(11), 1928.

[105] Thomson J. J. *Electricity and Matter,* Constable, London, 1904.

[106] Oraevskiy, A. N. Whispering gallery waves. *Quantum electronics;* 2002, 32, № 5, 377–400.

[107] Barber B. P. et al., *Physics Reports.* 281, 65, (1997).

[108] Taleyarkhan R. et al., Evidence for Nuclear Emissions During Acoustic Cavitation. *Science* 295 1868 (2002).

[109] Landsberg, G. S. *Optics.* Nauka: Moscow, 1976.

[110] Rizer Yu. P., 1980. *Foundations of modern physics of gas discharges.* – Moscow, Nauka, 1980.

[111] Kogelnik H. Theory of optical waveguides, In book : *Guided wave optoelectronics.* edited by T. Tamir (Springer-Verlag, Berlin, 1988).

[112] Spillane, S. M., Kippenberg, T. J., Vahala, K. J. *Nature;* 2002, 415, 621–623.

[113] Corum K. L., Corum J. F. Experiments on production of ball lightning by means of a high-voltage discharge and electrochemical fractal clusters. *Usp. Fiz Sci.* 160 no 4 (1990) 47-58.

[114] Oraevskiy A. N. Whispering gallery waves, *Quantum Electronics,* 32, no.5, 2002, 400

[115] Urutskoev L. I., Filippov D. V., *Physics-Uspekhi*, 53, No. 12 1355-1358 (2004).

[116] Avramenko R. F., V. I. Nikolaeva, L. P. Pokasheva, 2004. "Energy-intensive plasma formations initiated by an erosive discharge - laboratory analog of ball lightning." In *the collection Ball lightning in the laboratory.* M. Chemistry, 1994.

[117] Shoulders K., 2005. Charge Clusters, *Infinite Energy,* 61, 2005.

[118] Mesyats G. A., Ecton - an avalanche of electrons from metal, Phys.Usp. 38(6), 567, 1995.

[119] Lewis, E. H., 2003. The Ball Lightning State in ColdFusion, *Tenth International Conference, Cold Fusion,* USA, Massachusetts, Cambridge, August, 2003.

[120] Urutskoev L. I., Filippov D. V., 2004. Usp. Phys, 174, No. 12, 1355-1358.

[121] Vysotskii V. I., Kornilova A. A. Nuclear reactions and transmutation of isotopes in biological systems (background, current state, prospects) 2017. *International Journal of Unconventional Science* vol 17-18 (5) 34-42

[122] Akulov Yu A, Manikin B A. Helium isotope mass spectrometric method for studying beta decay of tritium (idea, experiment, application in nuclear and molecular physics), *Usp. Phys,* vol. 173 (2003), N11, 1187-1197.

About the Author

Vladimir Torchigin, PhD
Professor, Chief Researcher, Russian Academy of Sciences,
Federal Research Center "Computer Science and Control",
Moscow, Russia
Email: v_torchigin@mail.ru

Prof. Vladimir Torchigin, born in 1938, graduated from Moscow Institute of Physics and Technology in 1961. In 2002 he put forward the theory that the nature of Ball Lightning is optical, where a conventional white light that fell into a trap that it itself made for itself. All physical parameters and properties of such light have been derived from well-known laws of optics and physics. These properties are completely identical to mysterious and intriguing properties of natural Ball Lightning. At present about 200 scientific papers are published in leading scientific journals on physics and optics.

Index

A

B

C

E